OIL TITANS

OIL TITANS

National Oil Companies in the Middle East

VALÉRIE MARCEL

John V. Mitchell, *contributor*

CHATHAM HOUSE
London

BROOKINGS INSTITUTION PRESS
Washington, D.C.

Chatham House (the Royal Institute of International Affairs) is an independent body which promotes the rigorous study of international questions and does not express opinions of its own. The opinions expressed in this publication are the responsibility of the authors.

Chatham House, 10 St. James's Square, London SW1Y 4LE (www.chathamhouse.org.uk); charity registration no 208223

OIL TITANS: National Oil Companies in the Middle East may be ordered from:
BROOKINGS INSTITUTION PRESS
c/o HFS, P.O. Box 50370, Baltimore, MD 21211-4370
Tel.: 800/537-5487; 410/516-6956; Fax: 410/516-6998
Internet: www.brookings.edu

Library of Congress Cataloging-in-Publication data
Marcel, Valérie.
Oil titans : national oil companies in the Middle East / Valérie Marcel ; John V. Mitchell, contributor.
p. cm.
Includes bibliographical references (p.) and index.
ISBN-13: 978-0-8157-5474-9 (Brookings : cloth : alk. paper)
ISBN-10: 0-8157-5474-4 (Brookings : cloth : alk. paper)
ISBN-13: 978-0-8157-5473-2 (Brookings : paper : alk. paper)
ISBN-10: 0-8157-5473-6 (Brookings : paper : alk. paper)
1. Petroleum industry and trade—Government ownership—Middle East—Case studies. 2. Petroleum industry and trade—Government ownership—Algeria. 3. Government business enterprises—Middle East—Case studies. 4. Société nationale pour la recherche, la production, le transport, la transformation et la commercialisation des hydrocarbures (Algeria) I. Mitchell, John V. II. Title.
HD9578.M628M37 2006
338.7'66550956—dc22 2006004279

9 8 7 6 5 4 3 2 1

The paper used in this publication meets minimum requirements of the American National Standard for Information Sciences—Permanence of Paper for Printed Library Materials: ANSI Z39.48-1992.

Typeset in Minion and Univers Condensed

Composition by R. Lynn Rivenbark
Macon, Georgia

Printed by R. R. Donnelley
Harrisonburg, Virginia

Contents

Appendixes

List of Tables and Figures

Tables

Figures

Acknowledgments

This book is based on conversations with a large number of people in industry and government in the Middle East and North Africa, as well as Malaysia, Norway, France, the United Kingdom and the United States. I learned a great deal through so many frank and challenging exchanges. My first thanks go to all these people with whom I enjoyed fascinating discussions. I hope you find some of your thoughts and ideas in the pages of this book.

These visits would not have been possible without the support of some key people. I would like to give special thanks the following: Javad Yarjani, who so kindly organized my visit to Tehran and Isfahan; Waleed Al-Hashash for interesting conversations and introducing me to KPC; Mohamed Mazari-Boufares for explaining so much; Abdulla Nasser Al-Suwaidi for his kindness; and Hasan Al-Hasan, who juggled schedules so that I could meet every resident of Dhahran. I also want to thank Nadhmi Al-Nasr, a great friend, whose support was instrumental in preparing this book.

I would also like to thank the chief operating officers of these oil titans for allowing me to conduct interviews: their Excellencies Abdullah S. Jum'ah, Mehdi Mir-Moezi, Nader Sultan, Hani Hussein, Mohamed Meziane, Yousef Omair Bin Yousef, and Mohamed Johari Dasri. I am very grateful to the sponsors of the research: ADNOC, BP, ChevronTexaco, KPC, Statoil and Total. I will remember Jacques de Boisseson especially for his support and interest in the study, which kept me close to my deadlines.

Throughout my trips to these beautiful countries, I met many interesting women. I am grateful to those who kept me company in Dhahran, Shaybah, Isfahan and Tehran—Suha Shehri and Zahra Tarahomi Harandi, especially, whose company I greatly enjoyed.

The pages that follow were improved by the sound guidance of Robert Mabro and Willy Olsen. I am especially grateful to Mary-Ann Tétreault for her rich insight and her encouragement. I would also like to express my gratitude to colleagues at Chatham House for their help and support, and to Glada Lahn in particular for all her hard work over two years of the study. I am thankful to Chip Cummins, who read the book several times over, on planes and buses, always enthusiastic and constructively critical. I couldn't have written this book without John V. Mitchell. He has taught me a great deal about the oil and gas industry, economics and how to think differently.

VALÉRIE MARCEL
London, December 2005

OIL TITANS

Introduction

In most important oil-producing regions of the world, the oil industry has been nationalized. Ninety percent of the world's oil reserves are entrusted to state-owned companies. The five national oil companies (NOCs) that are the focus of this book together produce one quarter of the world's oil and hold one half of the world's oil and gas reserves. What do we know about these oil titans? Do we understand how they operate and what drives them? Do they emphasize politics over profits? Do they have the technical and business skills to develop responsibly the immense petroleum resources entrusted to them?

In spite of the dependence of importing countries' economic fortunes on the performance of the NOCs of the Middle East and North Africa, outsiders know very little about these organizations. Their opacity has discouraged most analysis so far. However, after explaining my project, I was invited to meet with the managers of Saudi Aramco, the Kuwait Petroleum Corporation (KPC), the National Iranian Oil Company (NIOC), Sonatrach (Algeria) and the Abu Dhabi National Oil Company (ADNOC). They allowed extensive interviews to be conducted, demonstrating their interest in discussing both their past accomplishments and future challenges. These were fascinating discussions, and this book will tell those companies' stories.

From the oil industry's perspective, it is an important time to be studying NOCs. Many of them are emerging on the international scene, no longer confined to filling the shoes of the oil majors, which left when the host countries nationalized their oil sector. NOCs have grand ambitions: they are competing with the majors on their turf by developing new oil reserves overseas and investing in international refining and retail activities. National oil companies from China, India and Malaysia are in the spotlight, signing deals in Saudi Arabia, Iran and Sudan, for instance. Some NOCs are being partially privatized, to encourage their competitiveness, while other countries are reasserting national control over their oil sector. In a move that startled the industry in the summer of 2005, the China National Offshore Oil Corporation (CNOOC), a Chinese state oil company listed on the New York Stock Exchange, bid to buy out a Western international oil company, Unocal—this bid presaged growing ambitions on the part of the new players. Meanwhile, Russia, which is seen as the new frontier for the private international oil companies, appears to be reasserting the role of the state in the oil and gas sector by dismantling the most successful private Russian company, Yukos, and discouraging new foreign investment. The industry is in flux.

Two years ago, when I began this research on national oil companies and to discuss them with peers, I received mostly negative comments about their inefficiency. But now, the tone has changed. People are starting to take notice of NOCs. And NOCs themselves are embolded by this change.

On the home front, national oil companies are often torn between national expectations that they should carry the flag and their own ambitions for commercial success, which might mean a degree of emancipation from the confines of a national agenda. They are not "just companies," however. Within their borders, they enjoy an unparalleled standing with the public. NOCs are often the best employer. They train young people to the highest standard, develop local technological capability, create opportunities for the private sector and develop the country's infrastructure. But more fundamentally, the historical context of their country's struggle for independence has made national oil companies politically sacred entities. As the former Secretary-General of OPEC stated: "Although several decades have passed since the era of nationalisation, our national oil companies continue to possess a rather unique political status in the eyes of their respective nations. They are still regarded as the symbol of national sovereignty that controls the most important and the most valuable resource endowment in our countries" (Subroto, 1994: 20).

Henri Madelin has explained the ideology of resource nationalism, which took hold of North Africa in the 1970s, as one which combines with a "revolutionary fervour directed towards removing all vestiges of the former subservience. It assumes a socialist color in order to be better designed to signify collective rights to interests accused of the three vices of being foreign, capitalist and private" (Madelin, 1975: 66). There remains a deep emotional attachment to sovereign control over natural resources among many Middle Eastern societies, as is also the case in Latin America. It follows from this that resisting foreign attempts to control oil is a tune that has more resonance in these countries than increasing the performance of the hydrocarbon sector. This might explain why no more in-depth analysis of these key actors has appeared in the Middle East than in consuming countries. People in producing countries should be debating, "Why have an NOC?" The answers would help to establish the priorities of the company—to maximize revenue, to promote national pride or to optimize the development of resources? This question is raised only rhetorically by those in consuming countries who want to see oil in the Middle East privatized. They favor a privatization of the oil and gas sector there because it would allow foreign oil companies to develop more oil for the world's consumers and market forces to overtake OPEC. In the consuming countries, few people ask, "*Why not* rely on NOCs to supply energy to markets?"

From the producing countries' perspective, a high-competence NOC with good governance processes may be preferable to a privatization of the industry that improves access to resources for foreign oil companies. NOCs are by statute, organization and often affinity more responsive to the needs of society and the interests of the state. Under optimal financial management, they can generate more revenue for the state than private companies, which have their shareholders to pay. In fact, in some cases they achieve this despite inefficiencies because the royalties, taxes and dividends they pay to the state are so high. They may also serve the state's foreign policy interests by supplying specific markets, and they are more willing than international oil companies to develop spare capacity in order to balance the world oil markets. Moreover, a number of their community investments have been designed to serve the long-term interests of society. This is the case of programs to train nationals and to develop capacity in the local private sector. However, a key challenge is to establish good governance processes within the NOC and between it and the state that will allow it to develop its competence and to serve national interests efficiently. Finally, some degree of competition from other companies can be a valuable means of stimulating the performance of national oil companies—though it may run into opposition from society.

In spite of the highly politicized national and international environment in which oil is produced and traded, the most relevant and urgent pressures on NOCs are economic and industrial rather than political. In fact, NOC managers are, for the most part, noticeably apolitical, perhaps with the exception of Iran. Their priorities are the allocation of resources for the industry, their operational autonomy and national economic challenges. Indeed, as companies operating primarily at home, they are preoccupied with the difficulty of meeting the needs of society and government as well as investment challenges to the industry. Contrary to a premise that guided my first steps in this research, I found that politics form only a backdrop to the operating environment of NOCs. This observation pointed to a complex relationship between NOCs, their society and their government that is essential to understanding national oil companies.

There are numerous divergences in perspective and interests between society, government and the national oil company. Events in the oil and gas industry carry a political significance for the societies of producing states. Popular political views will shape attitudes to, most notably, foreign investment. Nationals of the major exporting states will also expect oil export revenues to finance a number of public services. This public opinion pressure tends to be exerted on government and to reach the NOC only indirectly. Government is the interface between society and NOC. It is government's role to develop policies to achieve long-term goals, but it deals with contrary interests along the way and responds to short-term threats to political stability. This political process can lead governments to have incoherent goals—for instance, seeking greater revenue from the oil industry while asking the NOC to carry out social programs that hamper its operations and increase its costs. NOCs are instruments of the state and help government achieve a number of policy objectives. But NOCs' managements seek the freedom to manage the industry so as to mobilize the country's oil wealth without government intrusions that would cripple their economic and technical capacity. There is, in this respect, a constant institutional struggle between government and NOC.

This struggle is compounded by power asymmetries. The NOC is powerful because of its knowledge. It has technical and business expertise: it knows the fields and understands how the business works and what it costs. Government, for its part, sets the rules of the game: it determines the targets for the sector and decides whether to introduce competition and invite foreign investment. Society seeks information regarding the NOC's activities and influence over the government's decisions concerning the sector.

This book seeks to explain the complex relationship between the state, society and the NOC, and traces the industrial evolution of NOCs in order to identify new trends in their strategy and identity. National oil companies are evolving, seeking an elusive balance between their national and commercial missions.

Scope and Definitions

It is not my intention to set the national oil companies of the Middle East and North Africa apart from their peers. In terms of the historical importance of oil and the challenges faced by the NOCs in their industrial consolidation, the stories of these NOCs could be told alongside those of the Venezuelan, Mexican, Malaysian or Brazilian NOCs, for instance. NOCs worldwide share a common concern regarding their relations with the state. The geographic focus of this book is an attempt at comparing NOCs that are producers *within* a regional system. In itself, the focus on the Middle East and North Africa is an interesting one: the analysis will show that for all their common cultural, historical and political references, each NOC in this study is unique. It follows that it is not useful to create categories of NOCs. Instead, this book examines the way in which each company is unique and highlights differences between them.

The book focuses on companies with upstream activities because these are the revenue-generating activities that make them so important to their country, as opposed to the "downstream" or refining business, which historically has not been very profitable. Most Middle Eastern producers have very low production costs: some $2 for each barrel of crude oil that fetched around $50 per barrel in 2005 on the international markets. On the other hand, exploring for new reserves is an expensive and risky activity, though a necessary first step in producing oil. The upstream business is also highly charged politically: it is a central battleground of resource nationalism, which seeks to protect hydrocarbons from foreign hands. By contrast, the downstream business is not.

Moreover, in the minimalist definition[1] NOCs are not restricted to those companies owned entirely by government and in possession of exclusive rights over the mineral domain. A number of NOCs have evolved beyond the attributes that initially defined them and, to some extent, protected them. An increasing number of NOCs coexist and cooperate with foreign private companies on national soil. This is the case for ADNOC, Sonatrach, the National

1. I use a minimalist definition of an NOC: it is a company with at least 51 percent state ownership that is active in the exploration and production of hydrocarbons—in other words the "upstream."

Oil Corporation of Libya, Petronas of Malaysia, Petrobras in Brazil, Pertamina in Indonesia, NNPC (the Nigerian National Petroleum Corporation), the Qatar National Gas Company and PDVSA (Petróleos de Venezuela SA). Others have the structure of a private-sector company but with a majority of government shares at the voting level—this is the case for Petrobras of Brazil and Statoil of Norway. The configuration of the industry in Russia and China is still evolving but it includes both state and private-sector enterprises. With the growing commercial focus of most NOCs and the increasingly liberal agenda of governments legislating on these matters, there is more flexibility than in the past regarding the status of NOCs. In this more liberal view, they need not be exclusively state-owned operators of the national hydrocarbon sector. But there are forces within NOCs, state institutions and societies that resist this evolution.

Outline

A look at the history of the national oil companies in the Middle East and North Africa will help us to understand what they are now, how their present-day challenges have taken shape and how they perceive foreign investors and consumers. This history does not define these companies, nor does it determine their behavior; but it is in the background, a part of the producers' collective memory with which an observer of the region must be familiar so as better to understand conversations about the industry today. Chapter 1 therefore examines the history of oil concessions in the region and the subsequent nationalizations. Chapter 2 discusses current attitudes to this history and investigates how it has shaped national oil companies in the Middle East and North Africa. Chapter 3 introduces five national oil companies, the key characters of this book, by looking at their corporate culture, identity and processes.

The story shows that national oil company employees are proud of what has been accomplished since nationalization. But this is no longer a sufficient driver for success. NOCs seek to develop a strong commercial culture, and this permeates the aspirations of most managers. Saudi Aramco, for example, refers to the state as its "shareholder." The current trend in the Middle East and North Africa is one of emancipation from the model of NOCs as instruments of the state. There is a drive to adopt the business practices of private oil companies. Although these are clearly national companies with state ownership of capital, special status in the mineral domain, obligations to the national market and a common history, they say they want to operate like international oil companies (IOCs), and even talk about partial privatization. We will see, however, that the term "privatization" does not carry the same meaning everywhere. In the West, it usually refers to the process of engaging the private sector to provide services or facilities that are regarded as public-sector responsibilities or to shifting from publicly to privately produced goods and services.[2] In the Middle East and North Africa, this term has different meanings, both in different countries and in different sectors of activity. Moreover, even though many people in the industry showed interest in the management practices of private companies, the transition to a commercial culture is not fully implemented in most companies in the region. Many employees want to retain the community engagement that has historically defined their company.

NOCs operate in two very different settings: one is national, laden with obligations and constraints, but it is also where they draw their strength thanks to (sometimes exclusive) access to low-cost reserves; the other is the industry environment, which exists outside the rules of the national setting and in which they must be skilful to succeed. Chapters 4 to 7 look at how national oil companies manage the contending demands of these operating environments.

Chapter 4 focuses on the national setting, and examines the quality of the NOC–state relationship. Are the decision-making processes clear? Are the two parties' roles and responsibilities clearly demarcated? Is there political interference in companies' operational decisions?

Chapter 5 discusses the impact of demographic trends on national expectations regarding the oil sector and the revenues it generates for the state. National cultures are changing with the new generations, notably in respect of the work ethic and the degree of support that nationals expect from the state. Although societies in the region are changing, there are still enduring popular expectations that the state should provide essential services and employment to the population. Interestingly, this is not a matter of public discussion within the major Middle East and North African exporting countries. Nevertheless, this is to some extent part of the national political pact. The legitimacy of the political structure is tied to its oil policy and management of hydrocarbons, but even more directly to its distribution of oil revenues. Public attitudes and the political role of oil specific to each country shape oil policy in terms of exports, investments and the redistribution of wealth.

2. This definition is taken from Gordy Higgins, "A Review of Privatization Definitions, Options, and Capabilities," *http://rru.worldbank.org/PapersLinks/GlobalResults.aspx.*

The stability of these political systems depends on the capacity of the state to continue to generate sufficient oil rent to satisfy the needs of the population. It is in their critical role as wealth-generators that states depend on their national oil companies. (The weight of this political reliance on the growth of the oil rent is discussed in a special contribution by John V. Mitchell at the end of this volume.) In many oil- and gas-producing countries of the region, such as Saudi Arabia, Iran and Algeria, the needs of the population are growing faster than the oil rent. This analysis shows that these countries will need to expand oil and gas production in order to increase revenues but that even under relatively favorable oil revenue scenarios, the hydrocarbon sector will no longer be able to give the same degree of support to the rest of the economy for more than a few years. Governments will need to begin to show success in economic reforms, which have begun in most countries, in order to diversify their economies and reduce the dependence of the rest of the economy on the public sector and of the public sector on hydrocarbon revenues.

The national oil companies' mission will be to take on a share of responsibility for the welfare of the economy (by supporting the private sector, for instance), although success there depends on much wider diversification efforts. Chapter 6 addresses how the growing pressure on government to provide for the population is likely to affect its relationship with the NOC. We examine the evolving mission of the NOCs and how they are responding to growing needs by supporting the diversification of the economy while keeping an eye on their bottom line. Clarifying and improving the financial relationship between the state and the national oil company is important too. This relationship must cause the NOC to reduce costs but leave it sufficient funds to carry out programs of capital expenditure (NOCs' budgets are sometimes exposed to the short-term needs of government).

The NOCs were initially created as instruments of government policy, whose aim was primarily to assert sovereign rights over national resources. In a context of resource nationalism, they were to give the state control over the pace of exploitation and the pricing of its finite resources. NOCs also ensured that the state received an equitable share of profits. Since then, new challenges have periodically prompted changes in their structure and strategy. These challenges have stemmed both from the domestic environment (the national economy and the state's budget, in addition to the political aspirations of the country and leadership as discussed above) and from the external (industrial and geopolitical) environment. Chapter 7 considers the external challenges that make up the NOCs' international operating environment. These challenges include establishing new markets, keeping up with technological trends and responding to new environmental regulations. These are challenges common to all companies in the industry. The question is whether their national operating environment improves or reduces the national oil companies' capacity to respond to these external challenges.

Chapters 8 and 9 examine how the interplay between national constraints and industry drivers translates into the national oil companies' strategy. We will see that the NOCs have different strategic ambitions, both domestically and internationally. Chapter 10 focuses on the prospects for foreign investment in the region and the potential for partnerships. Just as NOCs are increasingly trying to penetrate international markets through upstream and downstream investment, so they often have to deal with increasing penetration of their own national market by international players. Internationalization and, conversely, the opening up of domestic markets to competition are changing the landscape of corporate oil. Chapters 9 and 10 give a glimpse of emerging trends, which point to an increasing blurring of the categories "IOC" and "NOC."

Methodology

Although NOCs play a central role in the economies of the Middle East and North Africa, it is quite striking that there is little scholarly interest in the mechanics and forces animating the oil industry. A survey of sources available in Arabic demonstrated a weak scientific and academic interest in the region's oil and gas industry.[3] There are very few books published on this subject. This is partly because articles are the preferred publication format of the region's academics; and in this format, oil is treated more often as a subject of political commentary than of industry analysis. The main concerns are geopolitics and geostrategy, war and U.S. interests. There is little analysis in Arabic on (nonpolitical) industry concerns. In English and French, the focus has also been on geopolitics and geostrategy since the Gulf war of 1991, which brought oil back into public view. The American and British invasion of Iraq in 2003 reignited the interest of French researchers and journalists in the geopolitical angle of oil, bringing their output closer to the Arab approach.

3. A survey was carried out for the purpose of this book by Dr. Karam Karam at the libraries of the Maison Méditerranéenne des Sciences de l'Homme and l'Institut de Recherches et d'Etudes sur le Monde Arabe et Musulman. Publications at the Center for Arab Unity Studies were reviewed as well. Internet resources in French and Arabic were consulted too. Please refer to the bibliography at the end of this volume for full details.

Whatever the language, there has been little interest in studying national oil companies, whose activities generate most of the regional governments' revenues—not to mention a large share of the world's oil production. NOCs work mostly behind the scenes. The aim of this book is to shed some light on the drivers, goals and challenges that they face and to point out some emerging trends of consequence for countries importing and exporting hydrocarbons. The lack of available data on national oil companies and the importance of understanding the producers' perspective prompted me to seek interviews with the national oil companies. These interviews are the book's primary source of information.

There are methodological limitations to relying on interviews. Facts and analysis vary from one interviewee to the next. To minimize the risk of giving too much weight to one interview, precautions were taken in analyzing the data. Consideration was given to the specific expertise of the interviewee—I would give different weight to the relevance of a refiner and a petroleum engineer talking about the upstream business, for example. Also, a large number of interviews were conducted in each company—between 14 and 36. Contentious statements reproduced in these pages, for instance a critical comment about the company, were confirmed by at least one other interviewee in the company. When diverging stories have been told, I reproduce both versions and constrast them in order to highlight the complexity of perceptions.

In addition, all companies were given an opportunity to respond to drafts of this text. Saudi Aramco, NIOC and Sonatrach offered comments on the analysis, many of which were incorporated in the book, and corrections of inaccuracies, which were carefully noted. The companies did not agree with all the statements made by their employees, but they did not object to the study being published on the understanding that these statements are not fact but perceptions of facts. This is also how the reader should consider what follows.

For all these methodological limitations and caveats, the interviews are an extraordinarily rich source of information. The picture one gets is nuanced and complex, which was particularly useful for this study given that much of the current information on NOCs in the Middle East and North Africa is incomplete or skewed ideologically. In fact, I had not initially planned to base this study so extensively on interviews, but this changed when I realized the eagerness of the national oil companies to tell their stories. I was surprised at their willingness to reveal as much as they did about their internal functioning and their perceptions of their environment. As a result, it seemed important to convey as much as possible of this new insight.

Indeed, in view of the reputed inaccessibility of NOCs in the Middle East, it came as a surprise that my requests for interviews with the companies and governments of Algeria, Saudi Arabia, Kuwait, Abu Dhabi and Iran were all greeted very favorably. I was given extensive access to all levels of management after an initial stage of explaining the study's purpose and that interviews would not be attributable. The significance of this access became apparent once I learned more about the NOCs: they are all engaged in a process of redefining themselves and attempting to find an appropriate balance in their relations with the state. This study was therefore timely, and many people in governments and the companies felt that the research would be a valuable external and independent contribution to an existing national debate on these issues. In both Iran and Algeria, for instance, controversial bills were recently introduced to their parliaments whose purpose is to redefine the relations between state and NOC.

Another reason for the receptiveness of the five NOCs to the study was their desire to be benchmarked. In this regard, they showed a great deal of interest in and curiosity about NOCs outside the region, notably Petronas (Malaysia), Statoil (Norway) and Petrobras (Brazil). As a result, additional interviews were conducted with the management of those NOCs. A number of respondents in the five NOCs also felt that they knew very little about how other NOCs operated and how they dealt with common challenges.

Some companies were more willing to open their doors than others. Algeria and Kuwait were quickly receptive to this research and to the idea of interviews within their ranks. One of their first concerns was how to benefit commercially from the findings of the study. Kuwait Petroleum Corporation's culture is one of open debate and self-examination. Algeria's Sonatrach is developing internationally and seeks greater exposure through specialized publications. Saudi Aramco, the National Iranian Oil Company and the Abu Dhabi National Oil Company took longer to convince. It is entirely justified that commercial entities should be cautious about accepting an outsider's request to collect data on their strategy and outlook. However, as I became more familiar with these companies, it became apparent that the initial reluctance to participate in the study was due to a specific corporate culture. This is changing, and these companies are communicating more and more with the outside.

NIOC and ADNOC have a more domestic focus than the other NOCs studied. For Iran, the national market is quite significant. Moreover, NIOC has not been exposed to international forums to the same degree as the others because it is operating under a U.S.-imposed embargo. ADNOC operates a complex system of consortia with private-sector international oil companies, mainly the original concessionaires, but the focus is on the technical challenges of developing the oil and gas. For Saudi Aramco, 2004 was a challenging year, because its estimates of reserves and production capacity were questioned by commentators outside the region. The company responded with new disclosures of data and an unusually bullish public relations effort. There is change under way within the company. Previously, it had built its reputation on concrete actions rather than words, in a strategy reminiscent of ExxonMobil's, and it also tended to leave public relations to the Ministry of Petroleum and Mineral Resources. But a number of its executives have been pushing for effective communication strategies, and the company has taken some courageous steps in that direction.

Interviews were conducted with approximately 120 people in 2004, primarily in the national oil companies of Saudi Arabia, Algeria, Kuwait, Abu Dhabi and Iran but also with the ministries of energy, planning and international affairs of these countries and with four regional banks. I met as well not only with managers from Petronas, Petrobras and Statoil, as noted above, but also with managers of Russia's Gazexport and executives at the private oil majors Total, BP, Shell, ChevronTexaco and ExxonMobil. Interviews were conducted under the "Chatham House rule": I could use the information I gathered but could not attribute it to a source. They were semidirective in character: as the interviewer, I would guide the discussion with a planned set of questions (though the selection and sequence of questions would be adapted to the discussion) but would allow the interviewee to answer freely. (The questionnaire is reproduced in appendix 1.) Those interviewed in the companies included the chief executive and his senior and mid-level deputies in such functions as corporate planning, marketing, refining, international affairs, distribution, petrochemicals, production, finance, exploration, new business development, research and development, human resources and public affairs. (The structure of each company is illustrated in appendix 2.) In interviewing employees from the most senior officer to the more junior managers, my intention was to gather a full range of perspectives from various generations and levels of seniority. Further, I spoke with four of the five oil ministers and with several ministry officials in each country.

The data from the interviews is supplemented by John Mitchell's macroeconomic analysis of the role of the petroleum sector in each national economy. This analysis is designed to emphasize in rough terms the dependence of the non-petroleum sector on the petroleum sector. A set of simulations of production, revenues and foreign exchange balances shows in broadbrush terms how long, and how far, these dependences can be sustained under various combinations of oil prices, production policies and adjustments in the non-petroleum economy.

Though past trends will be discussed in the first chapters, the objective of the book, as suggested by the chief executive officer of one of the NOCs, is to look forward and give a glimpse of future trends. Our forays into the past will therefore be limited; and throughout, they are designed to offer the historical context for emerging trends. As we shall see, new trends are carried by the post-oil boom generations, innovative industry partnerships, forward-looking strategies and policies and the ambitions of national oil companies.

one

How It All Started

Valérie Marcel and John V. Mitchell

Oil has played a seminal role in the historical emergence of the modern state in the Middle East and North Africa. For some countries, such as Iran and Algeria, this emergence is a story of colonial control by foreign oil companies supported diplomatically and militarily by imperial powers which needed oil to fuel their industrial economies. Conversely, it is a story of national emancipation in which the oil was taken back from the foreigners and used to support national economic development and to purchase allegiance to the new state. For other countries, such as Saudi Arabia, throwing off the direct foreign domination of the oil industry was less of an issue than controlling the influence of outside powers in the region and in the global oil business. Even in countries that did not fall under colonial rule after the collapse of the Ottoman Empire in 1918, the region's quasi-colonial status had one of several important effects. The emergence of aspiring, independent states in the Middle East was closely tied to asserting sovereignty over natural resources.

The world's great industrial and military powers, on the other hand, sought control of critical strategic resources as an advantage over their geopolitical and economic competitors. Control over oil resources in the Persian Gulf and North Africa was of supreme importance to France and Great Britain in the aftermath of the First World War and to the Allies in the Second World War.

Although the discovery of vast oil reserves in the Middle East made its producers the focus of attention, resource nationalism appeared first in Mexico. In a dramatic act, which virtually every Mexican supported, President Cárdenas expropriated in 1938 the holdings of most of the foreign oil companies in the country and created Pemex, the Mexican national oil company, to take over and run those companies' properties. Full of nationalistic fervor, Mexicans of all classes contributed money and valuables in order to help pay for the costs of the expropriation.

Similarly, when nationalists emerged as a driving force in the Middle East and North Africa, sovereign control over natural resources (their countries' only significant economic resource) took center stage in political emancipation, beginning with the nationalization of the Anglo-Iranian Oil Company in 1951. Foreign attempts to control the large oil-producing states, or at least to protect the status quo in the region, continued in the next decades, and external interference remains a catalyst of nationalistic feeling. Popular responses to the American and British invasion of Iraq in 2003 showed strong sensitivity to (a new round of) foreign powers' intrusion into the region's oil industry.

Public attitudes, official policies and industry action in the Middle East toward asserting control over national oil resources have evolved and changed over time. For many exporting countries, the overriding priority for three-quarters of the twentieth century was to gain or to consolidate the independence of both the state against foreign or former colonial governments and the national petroleum sector against foreign private-sector companies. The producing countries did not control oil production, sales and prices; these were locked in "concession" agreements with foreign private oil companies. This system would eventually be broken on the lines foreshadowed by OPEC's "Declaratory Statement of Petroleum Policy in Member Countries" of 1968 (see below). It justified an increasing state role in the petroleum sector, notably through the activities of state companies, as well as a tighter regulatory regime for access to exploration acreage and the setting of taxes and allowances. By 1980, the objectives of sovereignty and independence had largely been achieved, and national oil companies had been key instruments in realizing them. This chapter traces the broad lines of the emergence and evolution of the national oil companies, whose origins are from the time of the first concessions. Their history still lingers with, and sometimes weighs on, the producers of the Middle East and North Africa.

The Early Concessions

For the first half of the twentieth century, the international oil sector was dominated by a few American and European oil companies. They had developed from roots in two very different political soils. First, Standard Oil, Nobel, and the Rothschilds owned oil properties in the United States, Russia (pre-1917) and Romania (pre-1914) and they expanded overseas by exporting the oil produced in those countries through their highly concentrated marketing systems, pipelines and shipping networks. Second, unlike the above companies, the Anglo-Persian Oil Company and Royal Dutch (even after its alliance with Shell in 1907) did most of their business outside their home countries. They were founded by entrepreneurs who obtained large concessions in developing countries[1] and who also relied heavily on the support of their home governments.

Iran offers a striking illustration of the geopolitical backdrop to the concession era. The Shah of Persia granted the first Middle East oil concession in 1901 to William Knox D'Arcy, an Australian-British mining entrepreneur. D'Arcy had no company, only a secretary to handle his affairs (Yergin, 1991: 138); and when his funds ran out, he merged his interest with the Burmah Oil Company in 1905. Oil was struck in 1908. A year later, D'Arcy formed the Anglo-Persian Oil Company (renamed the Anglo-Iranian Oil Company Ltd in 1935 and British Petroleum in 1954). The British government was keen to support the concession, as this was a key to asserting its influence over southern Iran against the Russians and protecting the Indian Ocean against foreign imperial access by land. So interested was London in the geopolitics of the region that eventually it took a 51 percent interest in Anglo-Iranian.[2] Imperial pressures mounted, and the Anglo-Russian Agreement of 1907 divided Iran into Russian and British "spheres of influence," between which lay a "neutral zone" where the major oil discoveries were later made. A few British Indian troops were sent to protect the company's drilling sites against local tribesmen. By 1914, oil had become important enough for Britain to take measures to thwart regional ambitions toward its oil interests: upon the outbreak of war, it sent troops to protect the Abadan refinery against an Ottoman

1. A "concession agreement" is a legal instrument through which companies can conduct their oil exploration activities on a defined area of foreign territory. In the early days of the oil industry, the concessionaire was at liberty to manage exploration programs, production levels and prices.

2. The British government's authority was exercised by its nomination of two non-executive directors with power of veto over matters of national security but no special authority over commercial matters.

attack. The British government's interest was driven by Winston Churchill's perception that oil would play a major role in giving the United Kingdom a strategic and economic advantage over rivals.

Beyond Iran, the oil concessions map of the Middle East and North Africa also reflected colonial and imperial political, strategic and economic interests. Algeria had been firmly under the control of France since its invasion in 1830. Seepages of oil in northern Algeria had been known since ancient times, and leases were granted to French companies in 1877 and to British companies (registered in France) near Arzew. French interests in the country were maintained through state and private French companies participating in 16 upstream and pipeline companies, most of which had some foreign private-sector participation (Aïssaoui, 2001: 54–55).

The First World War had a decisive impact on the Middle East as a result of the defeat and disintegration of the Ottoman Empire. The pieces were distributed between France and Britain, except for Turkey, Yemen and what would become Saudi Arabia, which emerged as independent states. Under League of Nations mandates, France administered Syria and Lebanon while Britain administered Iraq, Jordan and Palestine. On the oil front, the British government, which was kingmaker in Iran, mandatory power in Iraq and protector of Kuwait and the Trucial states,[3] supported concessions to the Anglo-Persian Oil Company. The French government, trading some of its Syrian territorial interests, gained an interest for Compagnie Française des Pétroles (CFP), the ancestor of Total. It was formed in 1924 on that government's initiative but was wholly administered by private interests. The U.S. government came in from the sidelines under the slogan of the "Open Door" to secure similar treatment for its companies. The diplomatic offensive was launched in response to fears of fuel shortages and against what was perceived as patent British and French imperialism. It was successful notably because of the 1920 Mineral Leasing Act, which made foreign investment in U.S. public lands conditional on reciprocal treatment for American companies abroad.

Mesopotamian oil was the decisive battleground for rival British, French and American interests. Before the First World War, the British and German governments had agreed to reconcile their companies' competing interests in Iraq in a group, the Turkish Petroleum Company (TPC). This company was founded by Calouste Gulbenkian, an Armenian trader born in Istanbul who

3. The "Trucial states" were Arab sheikdoms on the southern coast of the Persian Gulf, which came under British influence by way of a nineteenth-century peace treaty.

became known as Mr. Five Percent because that was the share he won for bringing the parties together in the TPC. After the war, the French bought out the Germans, leaving no room for American companies. The postwar arrangements excluded them from Iraq and also from Iran, Kuwait and the other Gulf states under British protection. Washington challenged the validity of the TPC concession in Iraq; and in 1928, this transatlantic dispute led to the creation of the Iraq Petroleum Company (IPC), with 23.75 percent shareholdings each going to Anglo-Persian, Shell, CFP and an American group led by Standard Oil of New Jersey. Gulbenkian, a notoriously tough bargainer, retained his hard-fought 5 percent share. This arrangement was significant because all the companies committed themselves not to compete for concessions elsewhere in the former Ottoman Empire. The agreement was known as the "red line agreement." Though it set investment rules for oil throughout the region, it was created, like many borders in the post–First World War Middle East, by a simple stroke of the hand: Gulbenkian used a thick red pencil to draw a line along the former Ottoman borders.

The foreign governments and oil companies had thus not only decided the structure of the oil industry in Iraq but also created an explorers' cartel for the Arabian peninsula (except for Kuwait, which had been a British protectorate since 1899). The IPC cartel became the principal concessionaire in Bahrain, Abu Dhabi, Dubai, Yemen, Oman and Qatar, all countries under formal or informal British protection. In them, Britain used its political influence to promote large concessions, which favored the concentration of control into the hands of a few foreign companies rather than promoting competition between them (Philip, 1994: 42–43). Britain also insisted that the rulers of these countries grant exploration concessions only to companies under British control—until U.S. pressure caused it to withdraw the "nationality clause" in 1930 (Yergin, 1991: 283).

In inner Arabia, the political basis for the establishment of a foreign oil company presence was different from the rest of the Middle East. Ottoman rule of parts of the Arabian peninsula had not been taken over by another colonial power. Abdul Aziz Ibn Abdul Rahman Ibn Faisal Al-Saud, known as Ibn Saud, established and consolidated an independent kingdom through military efforts and dynastic alliances during the period 1914–30. During this period there was little foreign oil company interest in exploring for oil on the territory. This changed after the Kingdom of Saudi Arabia was formally created in 1932 (see below).

Elsewhere in the world too, in Latin America, where the United States was the dominant external power, oil concessions were controlled by a few foreign companies. America and European countries had often intervened in the domestic affairs of the independent ex-Spanish colonies; and for a time, "modernizing" governments in Venezuela and Mexico had allowed private property in subsurface minerals.

By the 1930s, the global map of concessions showed a network of cross-border company interests supported by a concentration of the oil industry. Competition between exporting countries could take place only through competition between the international oil companies. The companies contrived to prevent the development of an open international crude oil market, in which prices would have been driven down by increasing production in Iraq, Iran, Venezuela and elsewhere. In 1928, recognizing the problems of overproduction, overcapacity and the subsequent insecurity of markets, the leading oilmen of the day met to establish an "As-Is" understanding, allocating to each company a quota in various markets and setting a uniform selling price (Yergin, 1991: 264). This became known as the Achnacarry Agreement and it was intended to stabilize pricing and increase efficiency. The companies also concluded long-term sales of large volumes of crude oil at prices that were either fixed or benchmarked (Adelman, 1972: 92) and they determined the rate of investment in new capacity. As oil grew in importance in the exporting economies, this meant that the companies effectively determined the growth of the economy and the government budget: host governments could argue about but not decide prices and production. These companies came to be known as the Seven Sisters,[4] a name that still has resonance for producers. They had fixed the rules of the oil game, limiting the licensing, production and pricing options held by the governments of the producing countries. However, the cross-holdings of the companies stimulated contacts between the host governments, which was one of the steps toward the creation of OPEC in 1960.

Renegotiating the Terms of the Concessions: First Steps

The terms enjoyed by the concessionaires were challenged first in Latin America. Mexico, as mentioned above, expropriated most foreign companies

4. The Seven Sisters, the largest oil companies, were Standard Oil of New Jersey (American, future Exxon Corp.), Royal Dutch-Shell (Anglo-Dutch), the Anglo-Iranian Oil Company (English, later British Petroleum), Texaco (American), Socony-Mobil Oil (American), Gulf (American) and SOCAL-Standard Oil of California (American). (Gulf, SOCAL and Texaco later merged to create Chevron.) Sometimes an eighth sister was included: La Compagnie Française des Pétroles (CFP).

in 1938 and nationalized its oil industry. Venezuela took slower steps, renegotiating the terms of its concessions from 1942, with the support of the U.S. government. The terms obtained by the Venezuelan government established greater expectations for host governments. First, its hydrocarbons law of 1943 reasserted the state's ownership of subsurface rights: it could grant new concessions directly to foreign oil companies (between 1910 and 1943, the mining law had permitted concessions to Venezuelan nationals, who leased to foreign companies). The second element of this law, which was of greater significance to the Middle Eastern producers, allowed for the old concessions to be renewed for 40 years, but with an increase in the royalty rate for the government to one sixth. Further, the oil companies became liable for corporation tax, which was established for all sectors at 12 percent. Through double taxation agreements, the cost of the tax to the companies could be credited against their U.S. or other foreign tax liabilities. The important point was that Venezuela's sovereign right to set the rate of income tax was established. In effect, the relationship between the Venezuelan government and the companies was similar to that between the U.S. government and the companies which leased oil rights offshore or on federal land in the United States (Mommer, 2002: 113). The Venezuelan government used its sovereign power to increase the rate of corporate tax in 1947 so that the total "government take" of royalties and tax would always be fifty percent of the profits.

It was more difficult for Iran to improve the terms of its concession. In 1920, the British government pushed Iranian officials to accept an "interpretive agreement" which restricted the Iranian government's entitlement to profits arising in Iran and effectively excluded sales of Iranian oil abroad (Philip, 1994: 57). A year later, Britain sponsored the rise of General Mohammed Reza Khan, who deposed Ahmad Shah Qajar in 1925 and became the first Pahlavi shah, supposedly ruling under a European-type constitution (Farmanfarmaian, 1997: 45–46). In 1932, when falling oil prices depressed the Iranian government's revenue, Reza Khan abruptly cancelled the concession, before coming to an agreement in the following year that preserved the concession intact but on terms more favorable to Iran. The agreement of 1933 provided for increased royalties and a minimum guaranteed income while reducing the concession area by three quarters. In 1941, Iran was invaded by Britain and the USSR, there to stay until 1946. Nevertheless, emboldened by the examples it saw in Venezuela and Saudi Arabia (see below), the Iranian government demanded a 50-50 profit share. Negotiations with Anglo-Iranian broke down. In the turbulent political environment, in which the shah lost control of policy, the Anglo-Iranian concession was nationalized by the government of Dr.

Mosaddeq in 1951. In this, Iran was a step ahead of its neighbors, which undertook nationalization some 20 years later. In response to Mosaddeq's move, Anglo-Iranian prevented the sale of the nationalized oil wherever possible. The Iranians did not agree to submit the dispute to arbitration. As with Mexico, the U.S. government appears to have urged acceptance of nationalization with appropriate compensation, but the British government supported its company unequivocally (Philip, 1994: 119).

In 1953, however, Washington plotted with London to remove Mosaddeq and to reinstate the Shah. After the coup, Britain and America facilitated a renegotiation of the concession, with the British (BP's) share reduced to 40 percent and the remaining 60 percent taken up by U.S. companies. A crucial feature of the new arrangement was that the National Iranian Oil Company (NIOC), which had been formed in 1951 to run the nationalized operation, remained as the executive owner of the oil resources on behalf of the state. The foreign companies were "contractors" to NIOC.

Algeria remained under French rule until 1962. Because it was technically three departments of France, French mining law applied. However, France conceded a special petroleum code for the Sahara in 1958 that allowed Venezuelan-type concessions. Profits were shared 50-50, the concession areas were large and the concession periods were long.

Saudi Arabia, unlike Algeria, Iraq and Iran, was an independent country. But it was inside the "red line" and therefore out of bounds to competition between the IPC partners. And besides, they were pessimistic about oil prospects there. Keen to keep out competitors from that territory, however, they made an initial offer in 1932 which fell far below the one made by Standard Oil of California (SOCAL, later Chevron). In 1933 the king granted SOCAL a 60-year concession for eastern Saudi Arabia. Soon after, SOCAL brought Texaco (also outside the IPC) into the concession to form the company that would be known as the Arabian-American Oil Company (Aramco) from 1944. They struck oil in 1938. In 1947, the capital burden of expanding Saudi oil capacity led Aramco to seek to expand the partnership, but King Saud refused to admit European IPC members. As a result, the red line agreement was brought to an end in 1948 and the IPC was restructured in order to compensate the European companies for their exclusion from Aramco. The consortium in Saudi Arabia was enlarged to include Standard Oil of New Jersey (later Exxon) and Standard Oil of New York (later Mobil). On the basis of the Venezuelan example, the concession was then renegotiated to increase the royalties coming to the kingdom and to establish a 50-50 profit-sharing arrangement from 1950. In both Venezuela and Saudi Arabia, this profit sharing was effected mainly at the expense of the U.S. Treasury. A clever legal arrangement enabled profit accruing to the host government to be paid in taxes which could be offset against the companies' liabilities for corporate income tax at home.

In Kuwait, the original concession of 1934 was amended in 1951 so as to bring it into line with the 50-50 profit-sharing arrangement established between Saudi Arabia and Aramco in the previous year.

Fighting the Concessions Together

In parallel with these national developments, producers gradually set up mechanisms for concerted responses to the foreign companies' control of the oil market. The trigger for collaboration came in the last years of the 1950s, when the world oil market was depressed by the development of new oil sources in Algeria, Libya and Nigeria and the growth of Soviet oil exports. The oil companies responded by reducing their posted prices for crude oil, on which producers' royalties and taxes were calculated. After the first price cut, an Arab oil conference was held in Cairo in 1959, which the Venezuelan oil minister attended as an observer. In 1960, after Standard Oil cut the prices it was prepared to pay, government representatives of Iraq, Iran, Saudi Arabia, Kuwait and Venezuela met in Baghdad and formed OPEC, the Organization of Petroleum Exporting Countries.[5] In its early years, it attempted to form common attitudes toward oil prices, tax structures and allowances.[6]

The OPEC countries were not after just price control and greater petroleum revenues. Producers with mature oil industries and adequate government infrastructure wanted a say in how their most precious resource was disposed of. However, member governments' influence over production levels could only be indirect, because the concessionaires were responsible for the investment, production and marketing of the oil to which they had title. In fact, through most of the 1960s, oil demand expanded rapidly at the prevailing prices. Each exporting country's share of the market depended on the pressure it could bring on its concessionaires to maintain the ratio of their liftings from that country relative to the other exporters. The governments responded

5. Other countries, including Qatar, Indonesia and Libya, joined later.

6. According to the OPEC charter, adopted at the Caracas conference in January 1961, the organization's aim was to increase oil revenues for the member countries in order to contribute to their development; to assert their progressive control over oil production at the expense of the international oil companies; and to unify production policies, fixing production quotas if necessary for each member country.

to this dependence with OPEC's Declaratory Statement of Petroleum Policy in Member Countries of June 1968. This went beyond previous statements on prices, taxes and allowances and called for an increasing state role in the petroleum sector—directly through the activities of national oil companies and indirectly through participation in the ownership of concession-holding companies. Ideologically, the ground was prepared on the exporters' side for the nationalization and participation movement of the 1970s in which many of today's NOCs were created.

Changing Geopolitical Background

New forces were shaping the international system between 1945 and 1970 that ultimately created a political and ideological environment in which the great powers and their flagship oil companies could be challenged by the producing countries. Decolonization, the Cold War and the nonaligned movement were three trends that pointed to the decline of the multipolar world in which the imperial powers struggled to control resources outside their borders and to the emergence of a bipolar world in which the new states of the South could play the superpowers off against each other.

The process of decolonization was worldwide; and in rapid succession Britain, France and the Netherlands ceded independence to a large number of colonies between 1947 and 1957. In the Middle East, it was the decline of British power that was most dramatic. In Iraq, the British-founded Hashemite monarchy was overthrown in 1958. Kuwait became independent in 1961. Total British withdrawal from the Gulf was announced in 1968 and completed by 1971, at the end of which the United Arab Emirates was established in place of the Trucial states. Meanwhile, France was engaged in a civil war in Algeria from 1954 to 1962, having lost a colonial war in Indochina (1946–54).

For the United States, the geopolitical events outside the Middle East in the 1960s were profoundly disturbing and it consequently paid less attention to that region. Berlin was walled in by the USSR in 1961 and the Cuban missile crisis occurred in 1962. From 1964 to 1975, the U.S. was involved in a war in Vietnam, Laos and Cambodia. The USSR invaded Czechoslovakia in 1968, having invaded Hungary and suppressed a rising in East Germany in the 1950s. Under President Nixon, America initiated an "opening to China" in 1971 and concluded an anti-ballistic missile treaty with the Soviet Union in 1972. In the U.S., the peace movement focused on Vietnam; in Europe, it came to focus on the deployment of U.S. nuclear weapons.

With Britain and France losing ground and U.S. policymakers preoccupied by Soviet expansion around the globe, the Suez crisis of 1956 seemed to seal the decline of the imperial powers. Certainly, it was a watershed for the Middle East. Gamal Abdel Nasser became a nationalist, secular and regional hero after he nationalized the Suez Canal, Egypt's greatest source of foreign exchange and an important strategic asset. In response, Britain, France and Israel colluded in an adventure in which Israeli troops advanced to the canal and British and French troops landed, after some delay, in order to "protect" it as an international waterway. The U.S. refused to support the initiative, and there was a run on sterling. British and French troops withdrew without ever occupying the canal.

The Suez crisis had a profound effect on U.S.–Europe relations. The United Kingdom never again confronted the United States on a major foreign policy issue, and France gave priority to building the independence of Europe. Suez also demonstrated very clearly to nationalist movements in the Middle East that the power of Britain and France was gone and that nationalist states could successfully challenge any foreign power in the region by playing off the interests of the Soviet Union and the United States (Kissinger, 1994: 546). The U.S. could be expected to keep other foreign powers out but not to intervene against nationalism there except to the extent necessary to protect Israel from extinction.

In 1967, the Six-Day War provoked the first use of the "oil weapon." The governments of Saudi Arabia, Kuwait, Iraq, Libya and Algeria banned shipments of oil to the U.S. and the U.K. In response, the U.S. government lifted anti-trust restrictions in order to allow exchanges and cooperation between the companies that dominated the international oil trade; they were also the concessionaires in Algeria and the Middle East and controlled the loading and destination of oil exports. The companies just went elsewhere for their oil. Iran and Venezuela increased production, and within three months the embargoes were lifted. The lesson for the Arab oil exporting countries was patent: with the U.S. and European concessionaires in place, there was no effective oil weapon.

In sum, by 1970 the geopolitical context had changed in two important ways. First, British and French influence in the Middle East had diminished dramatically. Second, the U.S. was willing to accommodate nationalist policies rather than to allow the Soviets to extend their influence: this created opportunities for Third World countries to play off U.S. and Soviet interests. All this happened at a time of growing economic nationalism when, even in developed countries, the state sought to play a key role in managing industries and when governments in developing countries looked for state-led development—often at the expense of foreign investors. In this ideological context, the assertion of permanent sovereignty over resources accentuated three basic imperatives: to proclaim the integrity of the state; to reverse the unfair terms of trade that the West defended, in other words to change the rules of the game; and to speed economic growth by using the earnings gained from redistributing the take from resource exploitation (Clark, 1990: 160).

New Demands and Nationalizations, 1958–71

National oil companies in the developing world emerged either from nationalization, taking over the expropriated assets of the foreign oil companies, or from "participation" agreements in which the national oil company gradually filled the shoes of the foreign oil company as the state purchased the company's assets. Within these two types of NOC formation, there were variations or subtypes that distinguish each case. The key factors in NOC formation were whether the NOC had the time and mechanisms needed to learn or retain skills from the foreign oil companies and whether or not relations with the concessionaires and their home governments were conflictual.

The first wave of national oil companies came through nationalization. Mexico and Iran were ahead of a number of other producers in this process, even though foreign oil companies continued to operate in Iran, albeit as contractors to NIOC. Relations between the Shah's government and the foreign oil companies were difficult. The Shah constantly pressed the BP-led consortium to increase production, pay a higher take per barrel and invest in increasing capacity. In fact, Iran had increased its oil supply to the foreign consortium during both the Suez crisis and the Six-Day War. In 1966, the Shah threatened to turn to the Soviet Union for cheaper arms purchases, arguing that Iran's oil revenues were insufficient. The British government encouraged the consortium to settle, and terms were agreed. It increased production, supplied NIOC with discounted oil for bartering outside its members' own markets and reduced the area of the concession. In 1968, after Britain announced that it would withdraw all its forces from the Gulf, the Shah again demanded more revenue from the consortium, this time with the support of the U.S. government as well as the U.K. government. After a further confrontation, the consortium agreed (Bamberg, 1994: 172 ff). Foreign oil company activity and political interference in it continued until the revolution in 1979.

In Algeria, when independence was negotiated with France at Evian in 1962, the terms provided that the independent government would recognize previous concessions for 15 years, maintain the "Sahara code" for investors, give French companies preference in new concessions for a six-year period, establish a French-Algerian consultative body (l'Organisme Saharien) and submit to international arbitration those disputes that had previously been referred in the last resort to the French Conseil d'État.

However, after disputes over tariffs and taxes, the Algerian government sought to take a controlling interest in the group of companies developing the Trapal oil pipeline. French state and private-sector companies were involved in this pipeline, which was meant to increase the volume of the flow of oil from new discoveries at the giant Hassi Messaoud oil field in the Algerian Sahara. They claimed that Algeria's initiative would violate the Evian Accord of 1962, which provided that it would uphold the existing concessions regime. Despite strong resistance from the companies and the French authorities, the Algerian government created Sonatrach in 1963 as a state-owned company; it took 100 percent responsibility for the pipeline (Aïssaoui, 2001: 66).

More nationalizations were announced by the 1965 Boumédienne government. And in 1970, having first taken "control" (but not ownership) of U.S. and U.K. companies in Algeria as part of the oil boycotts in 1967, it nationalized their downstream and then their upstream operations. Finally, after failed negotiations in 1971, French interests were nationalized too. The concession system and the special French pricing and supply agreements were brought to an end. Despite these nationalizations, the American and French governments did not act to protect their companies' interests: both were concerned about the possibility of the newly independent Algerian government increasing its connections with the Soviet Union (Aïssaoui, 2001: Ch. 4).

In Kuwait, relations between the government and the oil companies became a topic of vigorous public debate during the 1960s. Though the terms of the concession had been renegotiated in 1951, the rising importance of Kuwait in the world oil market by the 1960s had caused some to question the existing arrangements. Oil became the dominant theme of the first session of the Kuwaiti parliament, established in 1963, when a group of opposition MPs criticized the practices of the Kuwait Oil Company (KOC, owned by BP and Gulf Oil) and the failure of the pro-Western government to combat them. In particular, it called attention to the overproduction of reserves, which had damaged the long-term recoverability of the Burgan oil field; to the practice of flaring gas associated with the production of oil, rather than putting it to productive use; to the limited progress made in training and employing

Kuwaitis in skilled positions; and to the capture of lucrative trade in supplying fuel to the Persian Gulf oil tanker traffic by KOC, which could have been carried out by the state-owned downstream company Kuwait National Petroleum Company (KNPC). A closely followed press campaign brought these issues to the attention of broader Kuwaiti society; and in parliament, the dispute came to a head over the Expensing of Royalties Agreement (established by OPEC in 1965), which renegotiated fiscal terms for the concessionaires on a basis opposition MPs saw as excessively lenient. The government stood firm about its close relations with the concessionaires, eventually passing the original bill in 1967 after calling new elections in which the opposition group was weakened (Al-Sabah, 1980: 27–32).

In Iraq, the monarchy founded by the British in 1925 was overthrown in 1958 by a coup supported by Nasserites and communists. The new government of General Qassim made a series of demands on the IPC; and in December 1961, it passed Law 80, expropriating without compensation all of the IPC concession except those fields that were actually producing. After General Qassim was overthrown in 1963 the nationalist Iraqi government gave the Iraqi National Oil Company (INOC) rights to all oil acreage in Iraq except the small area of existing production allowed to the IPC under Law 80. Later on the same year, it reorganized INOC and brought it under the direct control of the government. INOC adopted a policy of diversifying markets and development contracts, favoring especially the USSR and France, which supplied arms, aid and technology to the regime. In 1968, the period of political turbulence was ended by a Ba'athist coup, after which Iraq was governed by the Revolutionary Command Council. The remaining IPC oil fields were nationalized in 1971 and 1974, with compensation given, as in other OPEC countries, on the basis of the book value of the installations. The Iraqi and contemporaneous Algerian moves to end the concessions by nationalization were supported by OPEC resolutions and indications that OPEC members would try to frustrate any attempt by the companies to boycott Iraqi or Algerian oil.

In Libya, the government of King Idris was replaced by a revolutionary coup in 1969. Under Colonel Ghadaffi, local processes of government were subsumed in a combination of inspirational leadership and grassroots populism. Abroad, Ghadaffi sought leadership of the revolutionary "Arab nation," paying little attention to international legal processes. BP's interests were nationalized in 1971, purely as an act of foreign policy, in response to the British government's failure as it withdrew its troops from the Gulf to prevent the Shah from seizing the disputed islands of Abu Musa and the Tunbs. In parallel with its aggressive attitude toward renegotiating oil prices, Libya nationalized 51 percent of the interests of other foreign operators, starting with those least able to bear the complete loss of their investments or oil supplies from Libya.

The situation in Abu Dhabi was quite different. Owing to difficult geological conditions and a lack of natural harbors, prospecting took place later than elsewhere: the first major discoveries were made onshore in 1960 and offshore in 1962. By this time, a major influx of oil revenues into neighboring Qatar and Saudi Arabia had already caused a number of locals to migrate in search of jobs and a higher quality of life. Pressure mounted for a renegotiation of the concessions, which looked increasingly outdated, in order to enlarge revenues and fund investments in healthcare, education and infrastructure. In 1965, a new agreement brought royalties and taxation into line with the OPEC norm and reduced the size of the concession area to the producing fields. This arrangement freed the government to explore in the remaining areas or to invite bids from other companies to do so (Heard in Hollis, 1998: 42–44). When in 1966 Sheikh Zayed replaced his brother as ruler of Abu Dhabi, he put pressure on the foreign oil companies to increase exploration and production so that his ambitious modernization program could be paid for, but he maintained a cooperative attitude toward them that contrasted with the nationalist confrontations elsewhere in the Middle East.

Participation versus Nationalization

In 1972, the producers that had not yet nationalized their oil industry were heatedly discussing the relative advantages of participation versus nationalization. Governments hoped to gain control of the industry in order to resist foreign dominance—this was deeply resented by the population—but they feared the Seven Sisters. They were concerned that a country that nationalized the interests of such powerful international companies would then be excluded from the world oil market in an industry dominated by those few companies. With participation, however, a government took a share of a company's concession, rights, obligations and operations in its country; and this was seen as an acceptable compromise that avoided the uncertainties resulting from nationalization. The Participation Agreement, which came into effect on 1 January 1973, was signed by Saudi Arabia and Abu Dhabi and the major oil companies operating in those states.[7] It allowed the producing countries to

7. Those companies were Exxon, Atlantic Richfield, British Petroleum, Compagnie Française des Pétroles, Gulf Oil, Mobil, Partex, Shell, Standard Oil of California and Texaco.

take a 25 percent share in operations, which would rise each year by five percent until it reached 51 percent. Compensation would be paid to the companies, but only for the "adjusted book value" of the facilities, not for the oil resources, and the companies would assist in marketing the oil. Qatar, and later Nigeria, concluded similar agreements with their concessionaires.

But the outcome was in fact a mixture of both participation and nationalization, for soon after the 1973 agreement, Abu Dhabi negotiated a further agreement with the companies. This took effect in 1974, and in it the government acquired a 60 percent share in operations and the companies retained 40 percent. Saudi Arabia and Kuwait, for their part, would nationalize the foreign oil company interests completely: the former would do so over a period of years and the latter rapidly.

When the Kuwaiti government submitted the proposal for an initial 25 percent participation, the parliament (supported by public opinion and the press) was hawkish. It pressured the government to demand a greater initial share from the companies and commitments to reduce gas flaring. In 1971, Kuwaiti officials explained to Gulf Oil and British Petroleum that the government wished to avoid a collision course with parliament and that therefore the issue of gas utilization could not be ignored (Al-Sabah, 1980: 34). The companies' response did not satisfy parliament. They apparently underestimated its influence and that of public opinion on the government. Parliament rejected the Participation Agreement; and from 1974, the Kuwaiti government began to nationalize the upstream industry, acquiring a 60 percent participation in the Kuwait Oil Company. But this hybrid participation–nationalization deal came to an abrupt end in March 1975 when Kuwait announced that it was going to take over the remaining 40 percent.

The Saudi government gradually bought up the Arabian-American Oil Company (Aramco). It acquired a 100 percent participation interest by 1980; and in 1988, it established the wholly state-owned Saudi Arabian Oil Company (Saudi Aramco), thereby taking over Aramco's activities.

The Creation of National Oil Companies

The creation of national oil companies helped the producing states to nationalize oil by giving them the technical and organizational means to take over operations from the private companies when the time came. In most cases, there was a period during which the national oil company and the foreign companies overlapped. As we shall see, the quality and duration of relations between the foreign companies and national oil companies had a significant impact on the national industry later on. This was a period in which the national oil companies could learn the business. In particular, the slow transition of participation agreements allowed states to acquire the title to their resources and to develop in partnership with foreign oil companies the means to become operators of their own oil fields.

This process was especially fluid in Saudi Arabia, where Aramco was progressively nationalized. It was a smooth transition whereby the original company's expertise and organization were largely retained thanks to technical and marketing agreements with the foreign consortium members lasting until the mid-1980s. Interestingly, there was another national oil company in Saudi Arabia, initially created to develop the kingdom's mineral resources: this was Petromin, established in 1962 as a public corporation active in the downstream business. However, it failed to develop the skills and expertise necessary to develop into the main NOC, and the Saudi government chose instead to keep Aramco as the operator of the upstream and to "buy it back" from the foreign oil companies.

By contrast, the National Iranian Oil Company emerged in the turbulent politics of mid-twentieth-century Iran. NIOC was set up in 1948 as a first step in the nationalization process at a time when Anglo-Iranian still held concessions to most of Iran's known oil. Between nationalization in 1951 and the departure of the last element of the international consortium in 1979, the role of the foreign companies in Iran was repeatedly challenged and curtailed. In 1973, the Shah negotiated terms for a new 20-year agreement; this was based on an assumption of getting the financial equivalent of the 25 percent participation originally proposed by Saudi Arabia (see above). When the Saudis gained more favorable terms, the Shah increased his demands. He was "intent on assuring that he ended up with a better deal than Saudi Arabia. But for Iran, participation was irrelevant . . . Iran already owned the oil and facilities." The consortium operated the oil fields, and the Shah wanted more control over the industry (Yergin, 1991: 585). He got what he wanted: NIOC took over the consortium's assets, and the consortium, as the Oil Service Company, continued to manage operations and to lift oil under a five-year contract. In this context, relations between the national oil company and the foreign operators were not conducive to a transfer of skills to the NOC. Finally, after the revolution in 1979, the new Ministry of Petroleum cancelled all existing oil agreements and took control of oil and gas operations through NIOC, the National Iranian Gas Company (established in 1965) and the National Petrochemicals Company (established in 1964). By the end of 1979,

NIOC had set up the National Iranian Drilling Company, which began drilling and also to maintain 27 abandoned rigs.

In Algeria, the creation of the national oil company took place against a backdrop of France's efforts to preserve its advantageous position in the Algerian oil industry established at independence. French interests were being challenged by rising nationalism, manifested in demands for a greater share of revenues, and by disappointment with the performance of the French oil companies. The Algerian government was concerned at the decline in exploration by the concessionaires, which stemmed in part from their fear of growing state interventionism.

Sonatrach was established in 1963 in order to take control of a pipeline linking Saharan oil to the Mediterranean. Over the next seven years, it gradually took on responsibility for the various activities of the hydrocarbon sector in Algeria and built up its technical and managerial capacity in anticipation of nationalization. The expansion of its mission to the upstream began with the conclusion of new agreements between France and Algeria in 1965. These agreements established a 50-50 joint venture between Sonatrach and the French public company Sopefal for exploring and producing in a large tract of the Sahara. They also gave Algeria 50 percent participation in the French-controlled oil company SN Repal, leading to debates about whether to give the initiative to Sonatrach, as a fully state-owned company, or to SN Repal, as the vehicle for closer cooperation with France. The matter was resolved in 1969 in favor of Sonatrach. SN Repal was instructed to restrict its operations to two existing oil fields in preparation for eventual liquidation. This consolidated Sonatrach's position as the national champion in the run-up to full nationalization in 1970–71 (Aïssaoui, 2001: Chs. 2–3).

Like Algeria and Saudi Arabia, Kuwait initially created national companies that participated only in the non-core areas of the oil sector alongside foreign companies active in the core upstream activities. Between 1957 and 1963, Kuwait founded the Kuwait Oil Tankers Company, the Kuwait National Petroleum Company and the Petroleum Industries Company as joint stock companies with private shareholders. The core exploration and production company, the Kuwait Oil Company (KOC), established in 1934 by foreign oil companies, was acquired by the Kuwaiti government in two parts, in 1974 and 1975. Because of political pressure exerted by the Kuwaiti parliament, the government indicated that it did not want to maintain special links with BP and Gulf Oil. In response, BP and Gulf Oil representatives went to Kuwait City for talks and pointed out that there should be consideration for the old relationship. However, the Kuwaitis told them emphatically that this would not happen, and that "Kuwait intended to take over 100 percent, that it was a matter of sovereignty, and that the question was not open to debate" (Yergin, 1991: 647). The concessionaires claimed $2 billion in compensation, but they were given only $50 million.

The Abu Dhabi National Oil Company (ADNOC) was set up in 1971 to engage in all activities in the hydrocarbon sector. It would have a number of affiliates and subsidiaries. After the Participation Agreement, it took on the state's interest in the concessionaires' rights, obligations and operations. Unlike most Gulf countries, Abu Dhabi never claimed 100 percent ownership of the industry. In 1974, it took a 60 percent participation in onshore and offshore concessions. Today, most production is still carried out by the pre-1974 consortia in which ADNOC is the majority partner, but it does explore for new concessions and may undertake developments on its own initiative. The original tax structure was retained, with the foreign companies paying royalties and taxes (under terms less favorable than before 1974: effectively, they receive a fixed margin of approximately $1 per barrel produced rather than a share of the profits).

Types of National Oil Companies

Countries that nationalized or negotiated the purchase of foreign oil company assets created different conditions for the creation of their national oil company. Of those that nationalized the industry, some created a national oil company well ahead of the expropriation of upstream assets. This was the case in Algeria, which, as indicated, created Sonatrach initially to control the pipelines. Other assets were nationalized over a period of years, which gave it time to build up its managerial and technical capacity. Similarly in Iran, state ownership of resources was entrusted to NIOC, which oversaw the work of foreign oil companies until complete nationalization. Kuwaiti companies operated in Kuwait's non-core activities from 1957, but in the upstream the foreign oil companies operating KOC coexisted for only a short time with the national staff before full nationalization in 1974–75. Also, after 1975 KOC did not retain personnel from the foreign oil companies on its board or in its operations.

In these cases of expropriation, the national oil companies were created in a context of conflict with the concessionaires or the concessionaires' home country. And though Sonatrach and NIOC coexisted with the foreign oil companies for some years before their departure, the difficult history of political emancipation from foreign interference (whether plotting and coups in Iran or war in Algeria) did not create the conditions for skills transfer and the joint development of resources. After full nationalization in Algeria in 1971 and the revolution in Iran, the national oil companies operated mostly alone. Iran and Algeria were deeply affected by their difficult relations with Britain and France respectively. Their NOCs have tended to be more politicized and their societies have also been more sensitive to foreign investment. In Kuwait, the population was also mobilized against foreign investment, but events played out differently because the rulers, who were seen as being pro-Western and too soft on the concessionaires, remained in place (as opposed to in Iran). As a result, tensions were greater between the government and parliament than between the government and the concessionaires. Kuwait was also different from Algeria and Iran in that its relations with Britain were relatively peaceful, even though those with the concessionaires ended on a sour note. In Abu Dhabi and Saudi Arabia, relations with Britain and the United States were also positive; and this may explain why those national industries are less politicized than those in Iran and Algeria. The historical conflict between the Kuwaiti government and parliament certainly explains why members of parliament remain watchful of the government's management of the oil sector.

For countries such as Saudi Arabia and Abu Dhabi, which decided on participation agreements with a gradual nationalization of the oil industry, there were greater opportunities to learn skills. ADNOC in particular still operates with the original concessionaires through partnerships with them in most of its operating companies. Saudi Aramco too maintains positive relations with the concessionaires. For instance, its board still includes former CEOs of foreign oil companies. As a result, these national oil companies developed strong managerial and operational processes at an early stage.

Even where there were participation agreements, however, governments throughout the Middle East were ill informed up to the 1970s about even the most basic facts that would enable them to formulate their oil and financial policies. Although oil companies were compelled to give the necessary information to governments about their operations, they did so at a trickle during the first years of participation. Governments and national oil companies had no direct access to the data on their reserves from the companies, and relied on secondary sources or the industry press. For all the large amounts of raw data that were transmitted from the companies to the governments during the first years of participation, there was a total embargo on "interpretative" data. Producers came to mistrust the foreign oil companies, seeing them as secretive and arrogant (Heard in Hollis, 1998: 49–50).

Nationalizations Transformed the International Oil Trade

The nationalization programs of 1971 took place against a background of the renegotiation of price and tax concessions under the remaining concessions. The process was begun by Libya in 1965 with its unilateral revision of the price used for tax valuations and was continued by the Iranian confrontation in 1966. In 1970, Venezuela increased the profit tax to 60 percent and asserted the state's sovereign right to set this figure. At the same time, the Gulf OPEC countries began a series of negotiations with the concessionaires on prices and terms. The negotiation process ended in October 1973.

After the outbreak of the Yom Kippur war and the refusal of the international oil companies to agree to OPEC's price demands, the OPEC governments declared that in future, prices would be set unilaterally, and they announced an immediate price increase of 70 percent. At the same time, Arab oil producers began cutting production in response to America's intervention (diplomatic and with arms supplies) to defend Israel against the Egyptian invasion of Sinai. This was the first "oil shock"—the second (in 1979–80) would follow the Iranian revolution. It combined price increases with physical disruption, which, at least temporarily, caused shortages of oil in the U.S., Europe and Japan. It also precipitated the final collapse of the concession system.

For those OPEC producers that were not yet equipped with national upstream oil companies, their governments quickly created them, either through the acquisition of foreign interests or through nationalization. National oil companies were essential to this process of unilateral setting of export volumes and prices: it was their responsibility to keep oil production and exports going, even if at restricted levels. It was a critical test of the emerging national industries.

In spite of the limitations of the fledgling national oil companies, they transformed the structure of the international oil industry between 1966 and 1976 by increasing national control over it and enabling unilateral pricing by exporting governments. By replacing the concessionaires upstream, the national oil companies made it possible for their countries to gain control of their hydrocarbon reserves and production. However, the NOCs could not step into the concessionaires' places in refining and marketing in importing countries. Nor was this necessary. The nationalization and participation projects had completely changed the structure of the international oil trade. Their effect was to "de-integrate" what had previously been an internal oil trade carried out mainly within the integrated oil companies or between them.

Throughout the 1970s and 1980s, an increasing volume of crude oil was traded outside these channels, supplemented by new production (supported by the higher prices caused by the "oil shocks" of 1973–74 and 1979–80) from the North Sea and other non-OPEC suppliers. In many respects, international oil became another commodity, with a spot market, competitive pricing and futures markets. For the importing countries, "security of supply," originally gained by concessions and special relationships, now depended on the totality of supplies to the international market and the degree of competition within it. For the exporting countries, security of markets, that is, of export revenues, now depended on their ability to regulate competition between them in the face of much wider competition from other producers, the availability of fuels other than oil and the economics of fuel efficiency. The newly empowered oil exporting governments faced a new set of challenges.

Oil policymakers turned to attempting a degree of cooperation in pricing and, when that failed in the mid-1980s, in supply. OPEC's efforts were insufficient or too late to contain the effects of the economic crisis of 1997 in Asia, where slowing demand led to an oil price crash. In 1999, OPEC established ad hoc understandings with other major exporters (Mexico, Norway and Russia) for a wider curtailment of supply. The memory of 1997 faded somewhat as prices rebounded with Asian demand in 2003–04. And with the current high prices, cooperation among producers is no longer necessary. However, some producers are uncertain about how and when to invest in expanding production after decades of surplus capacity. Saudi Aramco announced in 2005 a huge production expansion program; but interviews in the Middle East revealed that despite exceptionally high prices since 2004 and forecasts of sustained demand, producers remain concerned about a price crash. Internally, NOCs are increasingly preoccupied with optimizing the development of their resources and supporting the state in efforts to promote the diversification of the economy. The main burden of diversification, however, lies outside the petroleum sector. The recent oil revenue windfall may not last long enough to maintain the growth of the exporting countries without a further expansion of petroleum production and eventually a lessening of dependence on oil revenues.

Governments' expectations regarding the role of the national oil company are also changing. The ideological context that led to the creation of many national oil companies in the 1960s and 1970s saw the state play a central role in the economy. Most oil-exporting countries now either are members of the WTO or are seeking accession to it and to an open, market-orientated international system of trading and investment. NOCs are confronted by a new global liberalization agenda in which the role of state enterprises in general is diminishing. Competition, rather than central planning, is becoming the objective of governments. NOCs are responding to these changes by focusing increasingly on their core business.

Conclusion

The historical conditions in which national oil companies were created continue to have an impact on their organization today. The geopolitical backdrop unfolding after the First World War (specifically the persistent interest of great powers in the region's oil resources) has shaped popular views about oil in the Middle East and North Africa. This fueled the ideological trends of the 1950s–1970s, which were instrumental in shaping the national petroleum sector. It created the conditions necessary for nationalization of the oil industry, but also led the governments of the time to build strong welfare states, often with the support of the NOCs. Also, in varying degrees, the private oil companies left behind skills, structures and corporate cultures. Depending on the nature of the regional governments' political ties with the foreign powers and their commercial relationship with the concessionaires, the industrial emergence of the NOCs was more or less supported by experienced private oil companies. Levels of competence vary, from Saudi Aramco, whose organizational processes and style clearly show the legacy of its previous American owners, to NIOC, whose difficult break with the concessionaires and intruding foreign powers (culminating in a revolution) made it hard to retain skills. NOCs' corporate cultures show telltale signs of their origins. As we shall see in the next chapter, this history continues to play an ambiguous role in shaping the worldview of NOC employees.

two

How History Is Viewed by the National Oil Companies

History for many of today's oil professionals in the Middle East and Algeria is the history of their country *since* its emancipation from foreign interests. The managers who started their careers in the national companies can tell the story of how they took over the industry, but usually they do not look back. As a result, the next generation, which joined the company when it was well established, does not know the story. And although students across the region learn the fundamentals of resource nationalism, important lessons about why their industry was nationalized are not taught sufficiently in schoolbooks, in the halls of OPEC, or when young professionals are integrated into the NOCs. This was apparent when I asked some of these young men and women why their oil industry had been nationalized—few could give a coherent answer (with exceptions in Algeria). In one instance, when I interviewed two Iranian engineers in their early twenties in front of two of their older colleagues, they hesitated to answer and turned nervously to their superiors, saying, "I'm sure Mr. ______ would be much better able to answer that question." The older men objected, saying the young engineers' answers did matter, and turned to listen with a little smile. They were curious to hear how the young men would describe the reasons for nationalization.

The first young engineer said, "There are two sides to it. First, you have the case in Norway where they found oil when the industry was mature [the country was industrialized]. Then you have the situation where we found oil in 1907. There was no industry. We needed the foreign oil companies to produce oil. Also we couldn't utilize much of that oil. Now we have a different industry. We know how to do it. I don't think right now we need a company-oriented oil industry. . . . We can hire contractors to do the job, which would be most commercial. Right now we don't need foreign oil companies to manage our oil industry. It works. It needs some modifications but it works."

The second engineer only answered that "Some reasons were political," and would not elaborate. It appeared to me that he had political views but that he did not feel comfortable expressing them in that setting, presumably neither in front of his superiors nor in front of a foreign visitor. His young colleague was more open, and picked up the conversation again, returning to the present and explaining that what really mattered was succeeding commercially.

These answers are quite different. In the first, the message is opaque, but appears to be steeped in today's debates about foreign investment. It seems to imply that oil was nationalized because the country developed the national industrial capacity to run the business, but it omits the difficult period in which the NOC had to build that national industrial capacity in the hydrocarbon sector and forgets that the national oil company itself was built out of nothing. The second answer highlights a reticence in the companies to pontificate about oil nationalism.

Familiarity with the history of the nationalization of the oil industry provides a starting point for assessing how the conditions that led the producers to nationalize have evolved. Are the factors that drove nationalization in the 1970s still as salient today?

Does Resource Nationalism Affect Decisionmaking?

The Importance of History

As shown in Chapter 1, the historical reasons for establishing national oil companies derived from the state's intention to assert national control over hydrocarbon resources, which meant control over decisions on production rates, over development and sales and thus over the generation of revenues for the state. NOCs would also enable governments to maximize their take, that is, the share of hydrocarbon revenues that the companies transferred to the state. After nationalization, their mission quickly expanded to take on a wide range of responsibilities (see chapter 6). Of interest here is whether the national oil companies and the governments of the Middle East and North Africa still refer to the time of nationalization to explain the NOCs' purpose and whether or not oil nationalism affects decisionmaking.

Clearly, past accomplishments are a source of pride for young employees. I asked a Westernized young employee at Sonatrach, "What makes you most proud: what the company has accomplished in the past or what it can do in the future?" He replied, "The past is very important. Our history is really very important."

In Algeria and Iran's national oil companies, the historical dimension is in sharper focus than in those of the Arab Gulf countries. For those two countries, the process of nationalization was difficult, and was played out with violent battles for independence in the background. In NIOC, Iranians display a feeling of pride in their industry and some resistance to foreign "intrusions" in it. Their general attitudes toward the different nationals doing business in Iran offer an intriguing view of historical perspectives. France is well regarded for standing by Iran during sanctions, while some Iranians convey strong negative feelings about the way they were treated by British oil professionals before nationalization. Of all the countries visited, Iran was probably the most aware of historical lessons and proudest of its emancipation from foreign influence in the oil sector. In Iran, unlike in other countries I visited, I was sometimes asked if I worked for BP. This was indicative of a more protective attitude toward the hydrocarbon sector. In Algeria, by contrast, attitudes toward France have softened to the extent that little distinction is made between French investors and other foreign groups.

In the Arab Gulf countries (apart from Iraq), the creation of the national oil company is described as having been smooth and easy. This was reflected in answers to my questions about nationalization in which managers patiently explained that they did not so much nationalize as "purchase" their companies commercially. In these countries' oil industries and oil ministries, I found surprisingly little appetite for discussions about the political characteristics of oil, even among the young.

Attitudes to Foreign Investment

In most of the producing countries discussed in this book, there is a gap between the national industry's concerns and those of society and its opinion leaders. The national media, popular opinion and other political institutions, such as the parliament, are often fiercer in resisting foreign involvement. In Kuwait, for instance, the government and the Kuwaiti Petroleum Corporation wished to invite foreign investment to help the NOC develop the northern oil fields. This project, known as Project Kuwait, would have allowed the first significant activity by international oil companies in Kuwait's upstream sector since nationalization in 1975. However, these plans have been thwarted by parliament since 1993. Several members of parliament are vehemently opposed to the project and to the company's justifications for inviting foreign investment. In Iran, NIOC and the energy ministry faced similar opposition from the Majlis (the parliament) and the Guardian Council concerning their efforts to find terms of investment that could attract foreign capital and technology to the oil sector. Nevertheless it is clear that NIOC managers would not go "backward," to a foreign-run industry. Ministry of Petroleum officials echoed this view.

In most of the region's countries, the impact of nationalization on oil professionals is apparent above all in their latent resentment of foreign involvement in the oil industry and their support of the government's refusal to give equity access to national reserves. Granting international oil companies the coveted right to "book reserves"[1] allowed Algeria and Libya to attract much-needed investment and enabled Abu Dhabi to keep investors highly committed, but equity stakes in other large reserve-holders in the region are simply ruled out. Because of the history of concessions and nationalization, Saudi Arabia, Iran and Kuwait, which together hold 42 percent of known global oil reserves and 54 percent of global NOC oil reserves, will not offer equity stakes in their oil.

The Need to Control Development

The refusal of the large producers to give equity access is well known, but another concern is less readily acknowledged by investors: these producers share a strong feeling that it is important to maintain control of the management of reservoirs. They are no longer concerned with preserving sovereignty over resources: they have been in charge of them long enough for the matter to be settled (unlike in popular sentiment). Their primary concern now is to remain in control of the industry, and particularly of reservoir management. This concern is the result of their or their neighbors' past experience of IOCs overproducing fields; it is also fed by contemporary cases. A number of upstream managers in the Gulf cited the case of Shell's management of Oman's reservoirs as an example of aggressive and unsuccessful reservoir management by a private company. In fairness, it should be noted that these upstream managers had no first-hand access to the geological data for this case. A similar conclusion is drawn from the past development of reservoirs in Libya and Kuwait by a wide range of leading international oil companies.

1. Companies listed on U.S., Canadian or European stock exchanges declare their "proved" reserves under various stock exchange regulations. Trends in these reserves are important to the stock markets' perceptions of the future income potential of the company.

A more general NOC observation regarding IOC reservoir management is made about the development of resources in the Gulf of Mexico. In this area, historically open to private companies and now geologically mature (that is, largely explored), even though the IOCs active there have had higher rates of recovery than in the Middle East, depletion rates have been much higher as well. This causes Middle Eastern producers concern that this aggressive development would also take place in their country should private investors be given free rein.

NOCs' concern about resource management stems from the fact that the region's professionals see their drivers as fundamentally different from those of the IOCs. They feel that the IOCs are thinking not of the long-term prosperity of their country but of shareholder expectations of returns in the next quarter. This common observation regarding the IOCs' rationale highlights a lack of trust in them, which is an obstacle to NOC–IOC partnerships.

Abu Dhabi's national oil professionals do not share this distrust of the motivations of foreign investors, however. Their industry was never fully nationalized, and ADNOC has worked with foreign partners since its creation. Furthermore, Abu Dhabi has set conditions for foreign investment that satisfy its concern to optimize the development of its oil fields and national control as well as the international oil companies' wish to book reserves. In this arrangement, its foreign partners are so pleased to be able to book reserves that they are willing to make concessions on the pace of development. A Total executive indicated that his company is able to book reserves in Abu Dhabi that account for approximately 15 percent of its entire reserves base—a situation that appears common to the other foreign partners. In addition, this model is favorable to the producer because each joint venture involves several partners, which allows ADNOC to use rivalries to its advantage and maximize its exposure to different views on the development of the reservoirs.

The preference of senior executives in ADNOC is clearly for partnerships that bring together several foreign companies. They view the alternative model, of a foreign partner as the sole operator, as risky because it involves reliance on one technology and one approach. One ADNOC executive explained, "There won't be one operating company in any area. It's better to have input from several companies." The following exchange with one of their foreign partners illustrates this point:

> VM: "Some NOC managers in the region seem to think that ADNOC isn't sufficiently controlling the management of its reservoirs."

> IOC [disagrees]: "Well, that's Oman. Shell was alone there talking to itself for the last 30 years. . . . Here they don't have a reservoir problem. The UAE has always respected [OPEC] quotas. Here they are one of the only ones (maybe the only one) that are able to raise their production without even flaring. With flaring, well then they can really boost production. . . . But we are playing the long-term game. We don't damage the fields. . . . But we do waste time."

The major producing countries in the Gulf want to see their fields developed so as to maximize production over 50 years or more, even if that means producing below their short-term potential. The following comments, by two professionals in Saudi Arabia and Kuwait respectively, indicate this long-term approach:

> Saudi: "Extending the life of a field is our first priority."
>
> Kuwaiti: "Fields will last that long if you invest in them over the long run. . . . We have important challenges ahead and we need IOCs. We want them but we also want full control over the operations. There's a perception that IOCs want to exploit our reserves. People think of the concessions."

The Ghost of Compensation

In Iran's oil industry, nationalism and the fear of renewed imperialism are more present than elsewhere in the region. Given the country's turbulent history and the central role played in it by oil, this is hardly surprising. Iranian oil professionals tell stories of what they feel was the outrageous compensation they had to pay the foreign oil companies after nationalization. To this day, negotiations with foreign investors are driven by a concern to protect the Iranian state from compensation demands. One manager involved in negotiations with IOCs explained, for instance, that IOCs wanted oil price exposure (that is, to own a share of the oil produced so that they could reap greater returns when prices are high, while accepting the risk of lower returns when prices are low). This was denied because the Iranians were concerned that the foreign companies could claim a value for future oil in the event of a compensation claim. The IOCs also wanted access to equity and the capacity to book reserves. The Iranian negotiators agreed to the companies booking their share of expected oil produced but refused to give entitlements to oil that had not yet been produced. Again, the concern was that in the event of an expropriation or if the companies left

Table 2-1. Revised Terms of the Iranian Buyback Scheme, 2004

Contract	*PSA*	*Iranian buyback*
Return	Direct share of gross production (usually 50 percent)	Fixed return on investment plus 15–17 percent profit
Duration	Around 30 years	Up to 25 years (five to seven years until 2004)
Incentives for the IOC	Potential for large profits, substantial period of control over oil field	Safe return on investment
Disincentives for the IOC	Investment costs may not be recovered if discovery is modest or prices decrease	Profits will not rise in line with a substantial discovery or oil price increase

because of security issues, Iran would have to compensate them for what "could have been" produced. The foreign companies, for their part, were concerned more about financial returns than about expropriation.

In the end, Iran came up with an unusual contractual solution, called buyback. The Iranian buyback scheme differed from the conventional production-sharing agreement (PSA) in its shorter time span and fixed rate of return on investment. Under a PSA, the contractor has about 30 years to explore, develop and operate a field, with profits from production being divided between the two parties. Until recently, buybacks specified a five- to seven-year exploration and development period after which operation of the field reverted to NIOC and the contractor's initial investment was reimbursed; there would be a fixed rate of return (15–17 percent) on profits. The short time spans and the lack of opportunity for control over production, and therefore profit, were unpopular with IOCs; and in 2004, Iran introduced revisions in an attempt to make buybacks more attractive (see table 2-1).

Nonetheless, a broader problem of misaligned expectations between the producer and the investors remains; and a senior figure at NIOC felt that because of political feelings regarding oil and fears of imperialism and exploitation by IOCs, it would be necessary to find new forms of contracts to respond to producer concerns.

Differences in Outlook between NOC, Government and Society

Political feelings regarding the development of the nation's oil resources are deeply entrenched in the societies of developing countries. While NOCs demonstrate more concern for maintaining control over operations, governments are torn between popular concern about repeating past history by inviting IOCs to undertake long-term development and their own desire to maximize their oil rent. In this respect, they would not always agree with the NOCs' careful and slow development of resources. A senior official in the Algerian Ministry of Energy and Mines felt that the NOCs were not given enough incentives and capital to invest in technology and exploration. "With partners, we see that the fields produce more. Otherwise NOCs sleep on their reserves."

A senior exploration and production director in the Arab Gulf acknowledged a difference of views: "The political leadership pressures the companies to develop their fields quickly and not to develop people. The government has to encourage the company to develop its people's skills and to develop its fields without a rush." An unspoken concern on the part of NOC managers is that the government's preference for a quicker generation of revenues will lead them to turn to IOCs, thereby challenging the NOCs' dominant position. This shows how far Middle Eastern governments have come from the time of nationalization, and brings us back to the question of why have an NOC. The NOCs' tendency to develop the oil resources slowly does not always suit governments—no more than it did when foreign companies controlled the pace of development. If governments want more rapid development in order to fill the state's coffers, they might turn to foreign investment as a way to complement and stimulate the activities of the NOC.

However, NOCs are not being fundamentally challenged in any of the regional countries: they are a source of pride for society and, more concretely, they allow national control over the oil industry. Although the ministries of petroleum are limited in their capacity to control the activities of the NOC because of a knowledge gap (on a technical field level, for instance), there would be an even greater divide between them and foreign oil companies. Without NOCs, governments would need to establish a special regulatory body to audit the performance of IOCs. Also, NOCs help to implement OPEC policies and can maintain spare capacity so as to stabilize markets. Investing in spare capacity is costly and rarely commercially driven. In Abu Dhabi, the NOC–IOC partnerships include contractual obligations for the private companies to invest in spare capacity. As one Emirati manager explained, "We are all working to increase capacity but the shareholders don't like idle capacity. We have a moral obligation to have spare capacity to supply the consumers."

Resentment of IOC Attitudes

In Abu Dhabi, relations with foreign companies are very positive; but elsewhere, distrust of IOCs can lead to tensions. Perceived attitudes among the oil majors contribute to the NOCs' resentment. A number of NOC managers feel that the IOCs have not appreciated the change that is in operation in the producing countries. This feeling was aptly conveyed by a Russian oil professional: "Many failure stories [of IOCs trying to invest] in China [are a result of] IOCs overestimating what they have to offer and how valuable it is to the counterpart. I'm convinced that most of the majors don't respond properly to the changes NOCs bring to [the] business environment. Interestingly enough, rarely are NOCs called competitors by majors; rather, they see them as potential targets for cooperation."

In varying degrees, the oil industry in the Middle Eastern countries and Algeria has evolved in isolation or with relative autonomy over the past 25 years. The result, as one oil professional explained, was that "The distance between us grew; and when they were invited to return to the Middle East, they didn't change their expectations [even though things were different]." The change in the region's nationalized industries was the result of their maturation. They started with a limited skills base, with an uneducated workforce but with large reserves and, in cases such as Saudi Aramco and ADNOC, with the management support of the private companies. Since then, they have expanded their core business in order to integrate their activities through the value chain, and now they are embarking on internationalization strategies whereby they strive to be as competitive as the IOCs.

When I asked employees of the national oil companies about their company's greatest accomplishment, their answers highlighted a pride in what has been accomplished since the early days of nationalization. Successes include (tangible proofs of) emancipation from initial dependence on the IOCs for skills through the education and professional training of nationals, leading to the nationalization of the NOC staff. The following comments show this pride in what has been achieved independently:

Algerian: "[Our greatest accomplishment was] to develop capacity from upstream to downstream with our own means."

Algerian: "Developing gas resources was a huge challenge. There was no market. No financial resources. It was very difficult, as you can imagine, for a recently nationalized NOC to find financing from the banks. But we did it."

Saudi: "The company took illiterates and turned them into world class professionals."

Attitudes to Privatization

According to the World Bank, privatization gained momentum as a worldwide trend in the early 1980s, with states relinquishing ownership of industrial and financial firms. A second wave of infrastructure privatization began in the late 1980s and peaked in the late 1990s. A third wave, in social sectors and core government administrative services, has yet to get worldwide traction.[2] Broadly speaking, the privatization of core public services remains a sensitive issue in the developing world and in particular in the Middle East and North Africa, where populations have come to expect government involvement. In the light of the even greater sensitivity regarding the role of private foreign investment in the oil and gas sector, I had expected all talk of privatization in this sector to generate animated opposition in discussions.

I asked NOC managers and ministry officials whether they thought it important that the state should control oil and whether they would consider the privatization of their oil and gas industry. Their answers went against the common assumption that national oil companies want to keep growing bigger without competition from the private sector and to gobble up new commercial opportunities. In fact, there is consistent talk in NOCs about the privatization of activities and about subsidiaries considered "non-core," whose assets could be sold to local investors. This is generally referred to as partial privatization.

KUFPEC (the Kuwait Foreign Petroleum Exploration Company), KPC's international upstream arm, has been one candidate for privatization. The idea behind the partial privatization of subsidiaries such as KUFPEC is to lessen the capital burden of the parent company and the state. This reasoning will become increasingly compelling when oil prices fall and when, in time, the government has greater difficulty in meeting the ever-increasing demands of the non-petroleum sector. Partial privatization also means that the company can maintain its national mission priorities, such as nationalization of personnel and outsourcing to local firms. Kuwait is unusual among the case studies in this book because a number of Kuwaitis support privatizing the oil industry as a result of perceptions that it is mismanaged by government (Tétreault, 2003: 80–82).

Abu Dhabi is considering privatizing some of ADNOC's companies, especially those providing services, in order to increase the competitiveness of the oil sector. The Iranian government would like to privatize the refineries, as

2. *http://rru.worldbank.org/Themes/Privatization/.*

would the Saudi and Algerian governments. But for this, they would all need to adjust the price-subsidy system, to make investment more attractive and to bring transparency to the accounting exchanges between the companies. Iran, like Saudi Arabia, will allow foreign direct investment in petrochemicals. The Algerian government has already taken steps in this direction. Saudi Aramco managers discussed the possibility of combining petrochemicals with refining assets so as to attract foreign investment. In Iran, oil ministry officials said that they would also like to invite foreign investment in liquefied natural gas and pipelines.

However, it remains very important that crude oil should be controlled by the state. As an outspoken Iranian manager put it, "If NOCs were private, the CEO of the NOC would be the leader of the country!" He was, of course, alluding to the excessive importance of oil to these Middle Eastern economies and polities. A Saudi executive commented, "Of course, oil has become so political that you can't touch ownership of oil. But if we were all starting over like the Iraqis, we might consider it rationally like the right thing to do."

Privatization is usually understood to mean selling state assets. In a number of cases, however, I was surprised to find that many in the Middle East understand it to mean achieving a more commercially orientated company; and in Iran, it was taken to mean decentralization. It was apparent that "privatization" did not carry the charged political meaning that I had assumed it would do in the region—perhaps because there is no question that core oil activities would be privatized and because for non-core activities, there is no assumption that privatization means foreign ownership. But it is doubtful that the region's private sector could weigh in heavily enough to compete with foreign capital. At an Iranian ministry, a senior official explained it thus, "There is not enough private sector in Algeria, Kuwait, Saudi Arabia, Iraq, Iran[,] etc. . . . what private sector?! The private sector in Iran doesn't mean billions; it means millions. We have small-size industry. We have 90,000 companies active in Iran, but less than 500 are more than medium size. When you talk about privatization in Iran, you're talking about the majors."

Recent Historical Events and Their Impact on the Oil and Gas Industry

The history of nationalization is by no means the only key to understanding attitudes in the Middle East and North Africa. Over the past three decades, significant events have had a defining impact on people's lives. In war-torn countries, there is a strong feeling of pride that comes from having developed the oil and gas industry independently and sacrificed so much to keep it going through sanctions and war. In Iran, successive upheavals have rocked the country for 50 years: independence and the nationalization of the oil industry, the Islamic revolution, the long war with Iraq and U.S. sanctions. These events involved extensive damage to the oil industry. Production fell from 5.7 million barrels per day (b/d) before the revolution to 1.3 million b/d two years later. Following the Iran-Iraq war, it took eight years for the industry to recover. Managers at a refinery in the imperial city of Isfahan told me about the courage of their colleagues who braved impending attacks by Iraqi warplanes to control fires in the installations. After the attacks, they repaired what they could, as best they could. The damage became a learning opportunity, as plant workers had to remove the nuts and bolts of the machines installed by the Western companies, open them up and try to fix them with the means available. Like the Iraqis, the Iranian oil industry people became self-reliant and resourceful professionals. Most available funds were needed for reconstruction in other sectors of the war-damaged economy, and there was little capital available for the oil and gas sector. The country was isolated regionally, the neighboring states having thrown in their lot with Iraq until 1990. U.S. sanctions have further isolated Iran, and the oil industry has found it difficult to gain access to technology and capital.

As Iranians often reminded me, Iran holds what it estimates are the world's top combined oil and gas reserves: it ranks second in each category. Therefore, they are acutely aware that their importance on the world energy scene is not in line with their reserves. Overall, it appeared from conversations on history and politics that Iranian oil professionals and officials were anxious for sanctions to be lifted and the industry to flourish in order for life in Iran to return to normal. Iranian youth generally displays strong curiosity about the West and a somewhat idealist view of the United States. They have lived in greater isolation than any of the societies discussed here (save the Iraqis) and have had little exposure to other cultures and perspectives. And yet, Iranian youth exhibit more pro-American views than their Arab neighbors, among whose students anti-Americanism holds more sway. A survey in Iran in September 2002 found that 74 percent of Iranians favored resumption of relations with the United States, even though national papers at that time dwelt on President Bush's labeling of Iran as a member of the "axis of evil" (Clawson, 2004: 16).

In Kuwait, the trauma and damage caused by the Iraqi invasion is still palpable. The country has not yet been completely rebuilt. Despite its wealth, a number of Kuwait City's apartment blocks are run down, their concrete facades crumbling. The war has also left scars on the oil industry, apparent in the poor state of reservoirs damaged by fires and the destruction of wellheads during the occupation and war. Restoration of the facilities and reservoirs to their previous state was partly halted by a combination of financial and managerial obstacles; and indeed, the country's and the industry's spending limits were themselves due in part to financial constraints resulting from the costs of the war. Kuwaitis told me that until the recent removal of Saddam Hussein, they had feared a new invasion and had put off some spending on reconstruction. In addition, some outsiders have commented that the sudden departure of Palestinians and Algerians from the country in 1990 left a managerial and skills gap in the Kuwait Petroleum Corporation that is still apparent. These Algerians and Palestinians, who held essential positions as mid-level managers and engineers after nationalization, were asked to leave the country when their governments supported or gave forms of support to Saddam Hussein after the invasion of Kuwait.

After liberation, however, Kuwaitis expected much, and people worked hard to rebuild the country. The newly reinstated parliament was determined to see things change after the war. It had been dissolved since 1985; and the new members of parliament, who tended to be better educated than in the past, wanted to have an impact on the management of the country, including the oil sector. A long-time observer of Kuwaiti politics and its energy scene, Mary-Ann Tétreault, has argued that there was a widespread belief among Kuwaitis that the government had failed in its duty to protect its citizens, notably by provoking Iraqi aggression with the overproduction of oil. In 1993, parliament, in its ambition to play a role in managing the oil sector, asked KPC for its vision of the future. But as time went on, things did not change as many Kuwaitis had hoped. Tétreault explains that Kuwait's rulers were slow to digest the implications of the invasion and Kuwaitis' thirst for change. The experience of those who remained in Kuwait and those who were abroad during the occupation contributed to a general awareness of the ability of private individuals and groups to achieve goals without state involvement. Many Kuwaitis were indeed less deferential to their government when it returned from exile, and the government did not make good on its promises to reform Kuwaiti politics (Tétreault, 2000: Ch. 5). It appears the loss of momentum demoralized some people in the oil industry as well. There is a perception that the ball stopped rolling at some point. In consequence, as we shall see in later chapters, civil society groups and members of parliament continue to be outspoken in Kuwait and do not necessarily trust government decisions, most notably those concerning the oil sector.

In Algeria, circumstances have clearly changed since nationalization. One manager's comment illustrates the themes that resonate in Algiers: "Nationalization was the logical result of independence. It fell into the logic of a struggle against imperialism. Now there has been a degree of political maturation. However, there is still the idea that we have to count on ourselves and make our own money by our own means. We must reap the benefits of globalization."

The Algerians interviewed want foreign investment, hoping that new opportunities will reach their shores. Algerians want a job, a house, a family—things one takes for granted in times of peace. Several employees expressed their relief at no longer worrying in the morning when they left their house about whether they would return at night. Their lives have been on hold since 1991 as the civil war raged and smouldered, and they are now emerging from a very difficult period with renewed optimism and a thirst for success. Fortunately, the Algerian oil and gas industry was not specifically targeted by Islamic groups in the civil war. In striking contrast to Iraq, where insurgents have relentlessly attacked oil and gas facilities since the beginning of the occupation, there is no need for rebuilding. The reconstruction of Algeria after the civil war is, in this respect, a social and economic effort. Algerians may be just a few years ahead of Iraqis in the struggle to return to normality. But the fact remains that both societies were robbed of the last 15 years.

Economically, Algeria faces huge challenges, and the extent to which it must open up to achieve this prosperity has been hotly debated. Globalization and the need to adapt to it came up in almost all conversations. The proposition that Algeria must expose its public institutions to foreign private competition in order to achieve prosperity is particularly contentious in the hydrocarbon sector. There is widespread concern that Algerian companies are not sufficiently competitive to survive "globalization within our borders" and that jobs will be lost. Consider the following excerpt from an interview in the Ministry of Energy and Mines:

> Ministry official: "Now, can I ask you one question? What do you think of globalization? There are many opponents of it. . . . "
>
> VM: "As a Québécoise, [I would support participating in globalization while keeping some state control of public services]."
>
> Ministry official: "Yes, but I want you to answer the question as an Algerian. What would you say as an Algerian?"

> VM: "I would say '*shwayya, shwayya*' (step by step)."
>
> Ministry official: "Yes, because we are a developing country. We are not Quebec. We cannot jump into globalization."

In other words, some forms of state intervention and state support of industry are still needed. Sonatrach will remain a national company.

In Saudi Arabia, the nationalization of Aramco was gradual and the Saudi side maintained positive and close relations with the American side. In fact, "Aramcons," as they call themselves, and oil ministry officials do not say that the company was "nationalized" but that it was "purchased commercially" by the government from the previous owners. On this point, a manager in his mid-forties commented, "The impression in Saudi Arabia was never that the companies took advantage and exploited our resources. Maybe there was even too much trust. After all, we didn't know much about oil."

Saudi Aramco still prefers to work with American companies, which it knows and whose disciplined managerial and operational processes it respects. For a long time, Aramco executives saw in ExxonMobil, their sister company, a model to emulate. However, ExxonMobil's reputation was damaged in Saudi Arabia by what Aramcons called its "political activities" during the first round of the Gas Initiative in the late 1990s. This initiative marked the first opening of the kingdom's upstream gas operations since the foreign companies left two decades before. But after the initial invitation to ExxonMobil and other companies made by the then Crown Prince Abdullah in September 1998, the negotiations dragged on. After the signature of a preliminary agreement in June 2001, failure in the several rounds of negotiations on the terms of investment led to the termination of the agreement, announced by the Saudi oil minister Ali Naimi in June 2003. Saudis told me that Lee Raymond, the CEO of ExxonMobil, would deal only with Crown Prince Abdullah himself, and declined to meet with Saudi Aramco. In the difficult negotiations that ensued, its executives explained, ExxonMobil tried to drive a wedge between their company and government through political manoeuvring. This hit a nerve, essentially because Saudi Aramco wanted to maintain the trust of the government. The negotiations were also difficult because they amplified existing political tensions and rivalries in the Saudi political system.[3]

Despite this negative experience, executives at Saudi Aramco hoped that other American companies will be involved in the kingdom. It is apparent,

3. Events surrounding the Saudi Gas Initiative are analyzed further in chapter 4.

however, that since 9/11, politics have been eroding the commercial relationship with America, and Saudi Arabia is opening doors to new partners. Regarding the difficulties Saudis have when visiting the United States, an exploration manager commented, "It's bound to have an impact. It's unfortunate for both sides. The U.S. has been very, very influential. It will lose that influence if it continues with its current policies."

The trauma of political turmoil is only now looming over Saudi Arabia, in contrast to Iran and Algeria, which have gone through successive political upheavals over the past 50 years. A terrorist attack was carried out in May 2004 in Al-Khobar, where employees of companies operating side by side with Saudi Aramco were brutally slain for being non-Muslim.[4] This was a traumatic event in a hitherto quiet environment. Employees are more concerned about their safety than before, and they wonder how long Saudi Aramco will remain isolated from turmoil in Saudi Arabia.

Abu Dhabi's modern history can hardly be termed turbulent. As part of the UAE, the emirate was ruled steadily by Sheikh Zayed bin Sultan Al-Nahyan until his death in November 2004. His leadership is accredited with the development and prosperity of the UAE. Sheikh Zayed, also the President of the UAE for over three decades, took a personal interest in the day-to-day affairs of the oil industry and made sure that oil revenues were invested in healthcare, education and national infrastructure. The political transition following the death of Sheikh Zayed was also very smooth, as Sheikh Khalifa bin Zayed Al-Nahyan succeeded his father on the throne without generating any opposition.

Conclusion

National oil companies look forward to the future more than back to history. However, historical events from the time of the concessions shape a number of attitudes, which may surface in their relations with international oil companies. NOC managements are proud of what has been accomplished nationally, and some do not trust IOCs to play a part in managing national hydrocarbon resources. Also, attitudes to issues such as foreign investment and privatization sometimes differ between NOCs, governments and their societies, because sensitivity to that episode in history varies. These differ-

4. I had the honor of meeting Michael Hamilton, a senior executive of Apicorp who fell victim to the attack. He is fondly remembered by his colleagues.

ences play out in more recent political developments too, such as in opposing views between the Kuwaiti government and parliament on how to manage the oil industry following liberation or in attitudes to globalization in Algeria. Each country has known specific political events, some recent and some past, that have played a role in defining the constraints and aspirations of these companies.

three
Corporate Culture and Identity

This chapter introduces the five main characters of the book, the national oil companies of Iran, Saudi Arabia, Kuwait, Abu Dhabi and Algeria, by examining how they identify themselves, their values, their management processes and their assets. Established companies have an identity and a specific culture that fashion the way employees operate and think about their business. Corporate culture is shaped by factors such as a company's structure and history and the culture of its country. This is all the more true for national oil companies, which are often a source of national pride because they embody aspirations of independence and success. However, as we shall see, these companies have been evolving, and change brings some internal questioning about what they are and where they should be going.

Though it is difficult to reduce these individual companies' culture and identity to a few words, we could summarize as follows: Saudi Aramco seeks to be the best, to surpass other NOCs; it gives a special emphasis to professionalism and technology. The Kuwait Petroleum Corporation and its subsidiaries will seize an opportunity quickly; they are adventurous investors. They also spend a great deal of time criticizing their organization and Kuwaiti politics. The National Iranian Oil Company's culture is to a large extent about sacrifice, for Iran and for the national industry, and about respect for colleagues. The Abu Dhabi National Oil Company and its operating companies emphasize respect too—of their national colleagues and their foreign partners equally. Their culture is consultative and non-confrontational. Sonatrach is an ambitious company with an eye to the world but most of its head still in Algeria.

Identity

When oil companies grow to be successful internationally, they become flagship companies. This was the experience of the British company Anglo-Iranian and the Italian company ENI, for example. For companies that operate nationally, internationalization is a graduation to a new corporate level and a source of renewed pride. Sensing the Middle Eastern and Algerian oil companies' simmering ambitions in view of the current trend of internationalizing NOCs from the developing world, I began my interviews with a question that was likely to hit a nerve and get people talking: "Would you describe your company as national or international?" This evoked strong but confused responses in most cases. The question is provocative in that it directly addresses a central preoccupation of many NOCs in the region: they are ambitious, increasingly international, and they do not want to be held back by their national status. They want to operate like IOCs, though they are clearly national companies with public ownership of capital, special status in the hydrocarbon domain, obligations to the national market and a common history. The oil professionals were fully aware of the changing profile of NOCs. Many interviewees felt that the industry leaders among fellow NOCs such as Statoil of Norway and Petrobras of Brazil, which are perceived to have extensive international operations, are harbingers of future trends for NOCs. These NOCs are challenging international oil companies on their own turf. As a result, there is a blurring of categories between IOCs and NOCs. The following comment by an NOC manager for new business development illustrates this: "The definitions are changing . . . An 'NOC' was a company owned by government and operating on the national territory. This is changing. There are no more NOCs. They all have international activities. We need new categories, a new name. Maybe 'public companies,' though that sounds like companies with public shares . . . 'Government-owned companies'?"

Sonatrach, in particular, has seen its historical special status in the national hydrocarbon domain challenged by a reform of the sector that has introduced competition, and it is building an international development strategy. This transition brings self-questioning, as a corporate planner explained: "We are still looking for our way. Where to go? Who to be? An international company or a national oil company?"

The Kuwait Petroleum Corporation also has a very international focus. It is aggressive in pursuing international investments both upstream and downstream, but it functions very much like an NOC in its operations and management processes. Saudi Aramco, by contrast, has a strong national focus, but it runs like a private oil company. Aramcons described their company as a hybrid: "not really an NOC or an IOC."

Iran was in a different situation altogether. Except at the highest levels of management, the company is not asking questions about its possible internationalization yet. It is so clearly a "national company" that this issue is irrelevant, and I quickly dropped it as an introductory question. In ADNOC too, this was not a preoccupation. International activities are generally left to its sister company International Petroleum Investment Company (IPIC), which invests internationally. As a senior executive commented, "We are not international except for marketing."

A Culture of the Heart or the Mind?

The NOCs under study have evolved in a unique set of circumstances that have deeply affected their corporate culture: they were created to become the national custodians of the most prized and political commodity of less-developed countries. But even though the companies' national role does make employees proud, past successes and the nationalist credo of the 1970s are no longer sufficient to sustain a sense of pride in the company. NOCs are developing a strong commercial culture, of profitability, innovation and competitiveness, and this permeates the aspirations of most employees. It was surprising how many interviewees referred to the state as the "shareholder" of the company. The NOCs liked to make analogies between their relations with the state and those of private companies with their shareholders. This was especially prevalent in the way people talked at Saudi Aramco. By contrast, it is noteworthy that KPC's managers tended to refer to the state as the "owner"—and not too cheerfully either. They seemed to express frustration at being tightly controlled by the state.

I asked NOC employees whether national oil companies in the Middle East should develop their own business style or strive to have the best Western-style business. I had hoped to elicit responses that would point to future trends towards a specifically NOC business style, one that successfully tapped into the strengths of the local culture and offered an alternative model to the Western multinational. However, it appeared that the current trend is one of emancipation from the NOC model and emulation of Western business practices. In the comments made, it appeared that working like an IOC means efficiency and high profitability and that working like an NOC implies operating like a government agency. Some felt that attachment to the national role was an obstacle to change: "The culture is that oil is a gift from God, that we all have a right to that gift of God. But who is looking at the technical costs in this culture?!"

The need to introduce a performance-driven business culture with the appropriate incentives and penalties was expressed many times. This was what most considered an IOC-style business culture to be, and many aspired to it as a vehicle for increased pride in the NOC. But the picture is more complex: these aspirations coexist with a strong attachment to the national oil company's historical role in providing training, welfare programs and opportunities for greater prosperity. Also, it is questionable whether they are willing to apply the actual necessary measures and working practices. For example, would NOC employees be willing to work the long hours of IOCs? Would they accept cost-cutting measures that involved firing employees? The following comment was made by a Saudi Aramco executive: "We inherited a blend of our corporate culture *and* our national culture. This blend is the Saudi Aramco culture. It's a heart culture more than a think culture. We don't want to hurt people."

NOCs are clearly torn between their commercial ambitions and their national obligations and cultural expectations regarding the work environment. What the Saudi Aramco executive called the "think culture" refers to business decisions made with what is often perceived in these companies as a harsh, numbers rationale that is guided only by profitability. The "heart culture" involves a concern for the impact of business decisions on the well-being of employees and society more broadly.

All young professionals interviewed wanted their company to remain the national oil company. Their attachment to the national status was not challenged by their international experience and education. But at the same time, these younger men and women wanted their company to perform, to be leaner, to be competitive. They would like to maintain their company's national status while introducing some Western-style management practices (though some did not identify these practices as Western). Fundamentally, they would like to see work practices improve within the national oil company model.

In this respect, young managers also exhibit an unresolved tension between their desire to emulate Western oil companies and their attachment to some aspects of the NOC's traditional business culture. Having learned the merits of contemporary oil company management processes, they would like to see these applied to their workplace. They would like their national oil company to run like an efficient machine, with dynamic, driven employees. As a Western-educated young Algerian explained, "[Sonatrach] should operate like the Western companies. It's very important for the employees' pride in the company to be on par [with IOCs]." It is apparent to most young managers, however, that the NOCs do not offer adequate incentives for performance. A number of people explained that performance is discouraged by excessive security and comfort: for example, employees get a bonus regardless of performance. One woman added, "A problem is that we don't know how to punish. There are no punishments if you don't do your job right." All those I spoke to also rejected the dominance of patronage networks and the use of *wasta* (influence through connections). "You need to appeal to personal relationships to make work progress and to get information from others. It shouldn't be like that."

But when confronted with the cold realities of a company applying these principles, these young managers feel deep conflict. Theirs is a society with social nets and family networks, and charity and generosity are strong values. Although many young people would like a meritocracy, they do not want it at all costs, as this comment from a female Saudi manager illustrates: "[We] aren't heartless. You don't fire people. Sometimes we think some people should go, but we think about their families, the people depending on them . . . I went to get my pants sown in a store and the old man working on the machines was so old that he couldn't see my pants and did a terrible job. But [at the] same time, I'm happy that we live in a country where a company won't fire an old man because he's not performing. We are not like Exxon-Mobil and companies like that who will fire thousands in a day."

Of course, the contrast between heartless companies that fire people and nice ones that do not is not really a private–public split. Private companies in Japan and the U.S., for instance, have at different times kept on personnel who were not productive. This was the ethic in private companies and in the public sector in the U.S. during the 1960s and 1970s and in Japan until the economic crisis of the late 1990s. Moreover, some interviewees challenged the notion that their companies really sought to foster solidarity: "They don't fire people, but they will put you on the side if they don't want you around." All these comments indicated that young professionals feel that hard work and innovation should be rewarded and "slackers" should be punished but that those unable to perform better, like the blind elderly tailor, should be protected from redundancy.

Company Culture

Most NOC employees want good management with a heart, that is, for it to apply the best business management practices while satisfying cultural expectations of generosity. National oil companies that have joined other private groups in partnership seem to have benefited from the mixing of business cultures. Saudi Aramco, for instance, matured in partnership with the American companies and gradually took over the reins of the company. This is also the case with ADNOC, which continues to work alongside the industry leaders. Because of its mixing of cultures, ADNOC is an interesting case. It is a company with a strong commercial culture and modern management processes but slow, traditional and consultative decisionmaking procedures. Indicative of its commercial culture, ADNOC managers and their IOC counterparts explained that the company punishes poor performance. This is highly unusual in the region as a whole. Another paradox is that this company is very hierarchical, even though every partner and every level of management is consulted in the decisionmaking process and encouraged to voice disagreements and new ideas. Furthermore, the company has transparent accounting practices and communicates freely with its partners and its "shareholder" the state, but it is wary of communicating with outsiders.

I also encountered this mixing of cultures in Equate, a petrochemical company in Kuwait that was created as a joint venture between KPC's petrochemicals subsidiary PIC, foreign private companies (principally Dow Chemical Company) and a Kuwaiti private-sector firm (Boubyan). These partnerships allow a transfer of Western-style business practices as well as technical expertise. Some of Equate's good practices have been learned and applied by PIC. A PIC executive explained: "[The joint venture] led to changes in all the processes. We are an environmentally certified company. We have a [human resources] plan. We are introducing IT for our billing (to have all bills go through the IT system and no paper). These changes are being carried through to the whole company."

It can be difficult to change the legacy of slow and cumbersome bureaucratic processes, which are often inherited from ties with government ministries. In Iran, for instance, NIOC and the Ministry of Petroleum have a symbiotic relationship, and, as a result, the company has developed a civil service culture. In Kuwait too, many of the parent company's personnel come from the oil ministry, bringing with them a bureaucratic management style. The bureaucratic culture is reflected notably in the lack of individual accountability and initiative. Even senior executives felt that this was a problem: "No one takes responsibility for anything. They always refer to the authority above them to get their approval for anything." And yet, in spite of the bureaucratic culture, employees of KPC and NIOC have a surprising capacity to be self-critical and to re-examine their strategy. For instance, when I discussed with individuals in those companies some weaknesses that others had noted in interviews or that I had perceived for myself about their organization, reactions were usually positive and open. They were not bound by a "party line" from higher management or by government directives. People talked freely about their company's limitations and were willing to re-examine their thinking and procedures. This is less true of Sonatrach, where managers have been sidelined for speaking up against plans to reform the hydrocarbons sector, and of Saudi Aramco and ADNOC, where employees have a strong sense of loyalty to the company. In these companies, individual initiative is not particularly encouraged—perhaps it is perceived as a lack of loyalty to the group, which is more valued in the region.

Each national oil company also has a specific culture that informs its management and decision-making procedures. KPC and its subsidiaries demonstrate a strong entrepreneurial spirit, and some of their foreign partners are taken aback by their willingness to take chances. Unlike many NOCs (and some IOCs), KPC is not a conservative company. Kuwaitis are also unusual in the regional context for speaking plainly and directly. One of the company's advantages that came up in discussions was Kuwait's merchant culture. This meant for some that Kuwait was a historical bridge between East and West. For others, it signified a tendency to develop deeper, long-term business relationships. Another feature of Kuwaiti culture that influences KPC is the holding of *diwaniyyas*. These are social gatherings hosted by leading Kuwaiti families, and sometimes by an industry group, as is the case with the Kuwait Oil Company. They are usually, though no longer exclusively, male. Rasha Al-Sabah, a cousin of the emir, has hosted a mixed *diwaniyya* since the mid-1980s. *Diwaniyyas* for women (often held in the mornings) flourished after the Gulf war of 1990–91 because women played an active role in the liberation and networked more freely than before. These were most often of a social or charitable nature, and the practice is now taking off among Kuwaiti businesswomen (Tétreault, 2005). These social events allow Kuwaitis to bypass the formal, hierarchical structure and gain direct access to others. In Kuwait, the small population and territory give Kuwaitis a strong and distinctive sense of proximity to each other. "*Diwaniyyas* are a good culture in Kuwait. There are no barriers. You have opportunities to go talk to anyone through the *diwaniyyas*. You know that all the big families have their *diwaniyyas* and [so do] some of the members of parliament. The prime minister has his own *diwan* and anyone can attend and talk to him directly."

Diwaniyyas are the channel through which *wasta* is exercised and special favors are granted. *Wasta* can help KPC by enabling its professionals to appeal to MPs in these informal networks and counter pressure from other, interfering MPs. The *diwaniyya* practice has strong potential applications to the extent that it allows informal relationships across the formal hierarchy. It is also a place in which to raise awareness of problems, to debate issues and to promote a better understanding of the needs of the oil industry. On the other hand, some favors are detrimental to the NOC because they are won through the *diwaniyya–wasta* channel and bypass regular management processes. For example, favors sometimes involve granting a job to someone who is not qualified. Thus it is important for KPC to insulate its management decisions from the social pressure of these networks. Consultation is also a central business feature in ADNOC, but consultations are held within the confines of the company and do not involve Emiratis outside it. The consultations are elaborate: there are committees for each sector and function (refining, marketing, technical, finance, and so on) and also board meetings for each company. The foreign partners are represented on both the committees and the boards, and at meetings all are encouraged to present their views before a decision is taken. This process fosters an atmosphere of cooperation, though it can be frustratingly slow for the foreign partners.

ADNOC's culture is shaped by its long-standing and close relationship with its foreign partners, including Total, Shell, BP, ExxonMobil and Japan Oil Development Company Ltd. (JODCO). To some extent, ADNOC plays the partners off against each other, and their competition stimulates the IOCs' incentive to transfer skills, knowledge and information to it. With concessions regularly up for renegotiation, the foreign companies also compete to impress the management of ADNOC. And yet, the atmosphere is surprisingly friendly. This is attributable in good measure to ADNOC's consensus-building approach, as made clear by the following exchange with an IOC executive:

> VM: "So the consensus-building process has the disadvantage of wasting time, but it has advantages as well because it builds trust between the partners?"
>
> IOC: "Yes, absolutely! All positions are open to all shareholders. But it's slow. Always slow . . . But in the end, there are more positives than negatives. A thousand times more positive. Essentially because our interactions are very courteous."

Sonatrach's culture is closely tied to the important role it plays in Algeria's economy and society. Several managers spoke of the company's "civic culture—*une entreprise citoyenne*." It has historically been a strong company, engaged in power struggles with government. Sonatrach appears to me like a young company awaking from a long rest during the turmoil of the civil war. Its employees want the company to succeed.

NIOC has not had the space in which to define a strong corporate culture. It is very closely intertwined with the Ministry of Petroleum. Nonetheless, some values emerge in discussions, such as sacrifice, independence and pride. The company's personnel are dedicated even though their salaries do not compete with those offered by IOCs in Iran.

Saudi Aramco operates in a controlled environment within a political and social system that is also tightly controlled. And yet it functions with its own rules within this system. It is, for practical purposes, exempt from a number of Saudi rules and laws, such as those relating to hiring women, to women driving cars, to men and women mingling, and so on. In addition, Saudi Aramco's culture is less formal and hierarchical than that of Saudi society at large—so remarked a young employee during a formal event at which the crown prince and his entourage were assembled in great pomp.

Employees obviously derive a great deal of pride from their company's social accomplishments, as the following excerpt from a conversation with a group of women (and one man) at the company illustrates:

> VM: "Do you think Saudi Aramco should continue to focus on its dual mission: the national mission as well as the commercial mission?"
>
> SA26: "Both. We don't want it to be a situation where we are not giving back to the community."
>
> SA20: "Saudi Aramco provides opportunities that wouldn't otherwise be there. Saudi Aramco is *in* Saudi Arabia. It's the country's main source of revenues. It's so important that it can't be just a company. It has a special status. Also, the money wouldn't come back to the community in the same way if it were given to the government for it to provide the services."
>
> SA23: "The company has so many infrastructure development projects."
>
> SA20: "It is the only company where women and men can sit together like here. Women drive in Saudi Aramco. It's like a small town. That makes Saudi Aramco special. . . . If Saudi Arabia changed, then there

could be IOCs here and Saudi Aramco could then be a simple money-maker."

SA24: "The company has a very human outreach. It is much more sophisticated than IOCs."

SA23: "It is even difficult for retirees to disconnect from the company."

Saudi Aramco is in the exceptional situation of a national oil company operating in a compound, like a foreign company. Though this presents the important advantage of insulating the hydrocarbon industry from political upheavals in the kingdom and also instills a formidable *esprit de corps* in the company, I felt that it carried the risk of the company developing what I called a "fortress culture" and resistance to change. During the interviews, I asked Aramcons what they thought of being a national company operating in isolation from the rest of Saudi society and whether they had a "fortress culture." Most were taken aback by this question, as it had not struck them before as strange to be in a compound, like a foreign company. One interviewee remarked, "The compound grew with nothing around it originally. There was no Dhahran. Now there are cities around us. . . . "

Saudi Aramco has long been known in the oil industry for its secrecy, but this is changing. Its leadership has sought to protect the space it has for operating independently. By not talking publicly, by not attracting attention and avoiding politics, the company can maintain the consensus in the government that there is to be no interference in its affairs. In this sense, it is careful to been seen as an operator, not a policymaker. Some managers now feel that this is a limitation on the company. A public relations manager wanted to see Saudi Aramco communicate more and hoped that I would mention to the top management my impressions of the company's secret culture. In fact, the communication transformation is already under way, guided by a number of rising executives who see a more public role for the company in the future.

The Role of Women in the National Oil Companies

I had not planned to focus on the role of women in the oil and gas industry, but once in the region I was struck by their presence throughout the companies. (For the percentage of women in regional NOCs, see table 3-1.) In Kuwait, women have been particularly successful. For instance, Seham Razzuqi held until very recently two very senior positions: she was Managing Director of Financial and Administrative Affairs and International Relations at KPC and

Table 3-1. The Percentage of Women Employed by NOCs in Saudi Arabia, Iran, Kuwait, Algeria, and Abu Dhabi, 2004

Oil company	*Percentage of women employed*
Saudi Aramco	2
NIOC	6 (Ministry of Petroleum)
KPC[a]	12
	51 (Ministry of Energy)
Sonatrach[b]	11 (of which 38 percent are managers and 3 percent senior executives)
ADNOC	n.a.

Sources: Interviews (2004–05); Kuwait Ministry of Planning (2002); Sonatrach Annual Report 2004.
a. The figures for KPC and the Kuwaiti Ministry of Energy are from 2001–02.
b. The Sonatrach figure includes only permanent employees of 100 percent Sonatrach-owned companies.

remains the OPEC governor. Sara Akbar was KOC's first woman petroleum engineer, and played a heroic role in the post-liberation oil well fire-fighting team. She later became Manager for New Business Development at KUFPEC (Kuwait Foreign Petroleum Exploration Company), which is the international upstream subsidiary of KPC. She has since founded her own oil company.

Kuwait is perhaps unique in the Gulf for giving women extensive access to education and career opportunities. However, a female manager explained that Kuwait's oil industry is less accessible to women than elsewhere in the Middle East outside the Gulf. KOC and KPC are reluctant to have women working in the field; they prefer them to work in the safer environment of headquarters. Nonetheless, my visits to Kuwait and also Algeria gave the impression that in both countries' NOCs and oil ministries, women work easily alongside men and enjoy equal status. In Algeria, 38 percent of women employed for Sonatrach work in executive positions. The Algerian Minister of Energy and Mines, Chakib Khelil, issued in January 2005 a directive for the promotion of female employment in the oil and gas sector, which reinforced previous commitments (MEM, 2005).

In Saudi Aramco, women have not so easily gained access to high-level management positions.[1] The proportion of women in the company is only two percent (it is naturally higher at company headquarters than in operations in the field). The company was very sensitive to this inequality. Aramcons are often trained in the United States, and their attitudes to working with women are more Westernized than those of most Saudis. When I visited Saudi Arabia

1. Women have nonetheless found interesting opportunities in Aramco. For example, Amal Al-Awami and Fatima Al-Awami are petroleum engineer specialists. Amira Mustafa is a supervisor of 30 geophysicists in the Exploration Department and Mae Mozaini, a second-generation Aramcon, is Head of Corporate Identity and Advertising.

in January 2004 to collect information for this study, it was inaugurating the Haradh gas plant. The preparations involved great fanfare and minute attention to detail because Crown Prince Abdullah would attend. In only a week, a great city of ornate tents was built in the desert area of Haradh in order to host the prince and some 3,000 guests. Rows of 200 palm trees were planted in this short time; grass and flowers strewn across the sand. There was some debate as to whether I could attend the event: I was the only woman and I was Western. Saudi Aramco's awareness of the need to present the kinder side of Saudi attitudes to women and to show its openness prevailed over protocol. I returned the gesture by wearing a strict Egyptian-style Islamic veil. I found myself in huge, airy tents surrounded by thousands of men wearing the traditional dishdasha and taking little notice of me. They were characteristically discreet. As I wandered through the site, I found that the organizers had purpose-built for my use a "lady room," an air-conditioned room with a number of toilets and washbasins, flowers and toiletries. When later I passed by the lady room, I found a queue of men waiting to use it. They scattered in all directions, smiling shamefacedly, as soon as they saw me. My hosts at Saudi Aramco were embarrassed that there were no female Aramco employees at the event, but they could not escape Saudi social conventions when outside their compound, especially not in the presence of the crown prince's entourage.

The social context of the kingdom is clearly a serious impediment in the matter of employing women. Women and men are not allowed to mingle, and it is an indication of the special status Saudi Aramco enjoys that so many women are employed in it and work alongside men. Companies operating in Saudi Arabia must normally request permission to hire a woman, and permission is usually denied—this was attested by some executives at Apicorp, the regional bank based in Al-Khobar. Even though Saudi Aramco is exempt from this requirement, it did request permission from the government before hiring a woman as Chief Reservoir Engineer, one of the highest positions an oil company can offer. Another female manager explained that after serving in this position for several years, this woman quit the company because she felt that she had hit the glass ceiling. The Aramco women I spoke to generally argued for slow change in favor of greater equality; they did not want to see abrupt change.

In the Iranian ministry of oil and NIOC, there are many women employees, and they are rising slowly through the ranks. They are taking up more than half of all university places and are beginning to outnumber men in the engineering schools. Several managers predicted that they would play a growing role in the oil industry and in the country generally. For now, however, women's professional opportunities remain somewhat limited by gender-restrictive work practices, though a growing number of women are working in management functions alongside men. For instance, the women I met at the Isfahan refinery worked easily with their male colleagues. But in some cases, it can be difficult for women to rise through the ranks or to attain positions in which they would work, debate and meet extensively with their male colleagues.

I found myself in a scene quite contrary to the event in Haradh when I was invited to stay for lunch at the Ministry of Petroleum. I was taken by a small group of women employees to a canteen exclusively for women. Thousands of black frocks and veils filled the large room. I alone was dressed in color in this sea of black; but again, no one stared. Friendly attentions were bestowed on me by women of all ages in Iran.

Data were not available on the percentage of women employed by ADNOC, though it appears to be rising, particularly in marketing, research and human resources. Women have not yet been admitted to the Institute of Petroleum for specialized training, however. Coeducational universities are a sensitive matter, and the company "is a bit conservative," according to a company engineer. Another problem, according to a local IOC executive, is that even though the company seeks to hire women, they tend to leave when they marry.

The important role of women in these national oil companies is attributable to how the NOCs are perceived in their own countries. Though IOCs have difficulty in recruiting skilled workers because in the West the oil industry is perceived as an industry of the past (as opposed to other high-tech industries), national oil companies in the Middle East and North Africa are the best national employers, and they are seen as an industry of the future. They attract female recruits because they can offer skilled positions to the large number of women who come out of engineering schools every year. They also offer women more attractive work environments than IOCs (and private companies generally) because employment is more secure and the hours of work are more family-friendly.

Management Processes

Decisionmaking in the national oil companies tends to be slow and careful. Managers want to protect themselves from the consequences of a bad decision; they ask consultants to assist in choosing the appropriate course or to prepare reports validating their decision. This cautious decisionmaking process is most apparent in the Abu Dhabi National Oil Company, where each level will make recommendations that are usually referred up the hierarchy before a decision is taken. It is in contrast to some of the smaller private companies I visited in the region, which empower their executives to make decisions independently. For instance, at Equate many decisions can be made at the mid-management level.

This careful culture is compounded in some cases by bureaucratic approval processes. The slow processes in some NOCs and in the relevant government institutions mean that projects may not come onstream in a timely fashion, owing to the time lag between the decision to invest and its implementation. When important investment decisions require a significant departure from previous strategy, it is particularly difficult for the whole machinery of the NOC and its government watchdogs to make a commitment to the new course. Even for routine capital expenditure, there are worrying examples of slow bureaucratic processes: one NOC had to wait a year to get approval to purchase new computers, by which time the models were out of date.

In Kuwait, some of KPC's subsidiaries complained that the parent company imposed slow processes on their spending requests, with the KPC board having to approve requests already sanctioned in the subsidiary's budget. But other NOCs appear to demonstrate more simplified management processes. A surprising case is the National Iranian Oil Company, where managers at refineries explained the ease with which they make decisions and communicate them within the company. The organization itself is a maze, but internal decisionmaking processes are relatively uncomplicated.

At the Kuwait Petroleum Corporation, the greatest challenge was related to the very functioning of the company. There was a pervasive feeling of frustration about the lack of implementation of decisions. This problem was attributable in part to interference from government and parliament and to the politicization of the industry. It was also related to management problems, such as insufficient conviction and follow-through; a lack of concern for quality; and poor communication (between the subsidiaries and the state institutions, between the subsidiaries and KPC and between the subsidiaries themselves).

In several countries NOC managers expressed frustration at the lack of instruments with which to motivate their personnel. Often interviewees complained that there were carrots available but no sticks. The practice of forgiving poor performance is damaging to morale in the companies because the active employees work among those who are not as energetic, as this comment from a KPC director shows: "Good, trained [nationals] can be further trained at the company level but they arrive in a company where there is total comfort. You can see that it's not a lean company."

With the exception of ADNOC, the NOCs do not fire poorly performing personnel. They will isolate them so that they do not harm the business or treat them poorly in the hope that they will leave the company of their own free will. I was told about the existence of shadow offices, sometimes a whole wing or floor of a company building, that keep these underperforming undesirables away from important business. No one checks what they do with their time or whether they come in to work at all. Some employees felt that it was a sign that the company had a heart because it was aware that the "bad employees" too had families to feed. Others felt that the practice was actually unkind, even ruthless. "Internal exile" is also a punishment for those who have antagonized the management by their strong difference of opinion.

At Saudi Aramco, respondents did not complain about the internal functioning of the company. That would run counter to its culture: there is a great deal of cohesion among the ranks and pride in the company. Also, it is clear that Saudi Aramco runs a tight ship and that therefore there is less to complain about. There are, however, some processes, related to the company's centralized and hierarchical management structure, that may hinder its adaptability to new external challenges. Managers tend to micromanage because they are concerned about any wrongdoings or inefficiencies being uncovered under their watch. Approval from superiors is sought for any deviations from usual practices.

Communications with the outside are particularly sensitive, as they have not been the norm in the past. Saudi Aramco is secretive but is now beginning to communicate. When asked why this was, a reservoir engineer explained that the company felt underappreciated and wanted "people to see all the things we do." There is indeed a great deal to show off about, but the company is finding it difficult to deal with this transition. Though it has made itself much more accessible to journalists and visitors in the past few years, it remains uneasy with media forces that it cannot control and that may not give a positive image of it.

Sonatrach has not opened up to the outside as much or as rapidly as Saudi Aramco did in 2004 and 2005. It is more forthcoming than many peers about divulging its financial and technical data, and certainly its management was quick to participate in this study. But it was quite sensitive about conclusions of the study that pointed out weaknesses both of the company and in its relationship with the state.

Regarding the company's management processes, interviews in Sonatrach suggested its employees felt that significant improvements to its structure had been made in 2000. The company is adapting a number of its IT, human resources and health, safety and environmental systems to international standards. In written comments, the company highlighted its innovative employment exchange, which identifies employees' competences and capitalizes on their skills by encouraging intracompany mobility. It also mentioned its new two-step bidding procedure for contracts, which has brought savings of time and money. Moreover, it regularly holds brainstorming meetings for managers in order to encourage innovative thinking. On the whole, it appears that Sonatrach has already identified and responded to a number of problems related to its internal management processes. However, a corporate planner felt that this process of structural reorganization needed to be a permanent one and that the company had to train its managers to raise productivity and reduce costs.

Internal processes at ADNOC appear to be clear and effective, though they are slow because of the company's consultative method, as discussed above. There were no complaints from employees on this front. The company gives incentives for performance and reprimands failures.

As national oil companies seek to succeed commercially at both the domestic and international levels, there is a great deal of appetite for implementing efficient, private company–style management processes, such as performance incentives. Also, a number of companies explained that their interest in participating in this study was motivated by a desire to be benchmarked against their peers. Benchmarking is seen as a potential driver for increasing productivity. Moreover, NOCs (Saudi Aramco excepted) are keen to learn best practice from the private industry leaders in order to run their operations more efficiently and effectively. Most require IOC assistance to integrate subsurface and on-field management. As a Kuwaiti exploration and production manager put it, "We are better at that now, but we still need to develop that more. IOCs are good on that—integration. That's their edge. For an investment they have to calculate their return and therefore they are more efficient."

NOC Strengths

Many Middle Eastern NOC executives admired Petronas of Malaysia as the new breed of NOC because it has extensive international operations. These executives deemed Petronas and other national industry leaders to be successful because they have made successful transitions to working like IOCs.

But they often ignored the fact that internationalizing NOCs have capitalized on their national status when venturing abroad. For example, the Malaysian government supports the international activities of Petronas through high-level visits and trade missions. In interviews, Petronas executives were not shy about acknowledging this political support and its value, but they also emphasized that the commercial terms get the deal in the end. A key to their success is the combination of private company–style management processes with their national oil company assets. These assets include long-term thinking, which means that the company will be more patient in seeing through the consequences of its investments and that it has time to look at softer issues in investments. Petronas sees itself as a hybrid—neither a traditional national oil company nor an international oil company. But its outlook is very much international.

Middle Eastern NOCs might admire this new breed of NOC, but they have not taken up its formula for success. With the exception of Sonatrach, the regional NOCs are reluctant to play on their national status in order to win bids outside the country—even in countries where they recognize that they may have an advantage in doing so. There is a marked preference among them to play on the same field as the IOCs and to demonstrate their equal competence. NOCs want to be seen by the NOCs and governments of host countries as international and commercial, not as national and political. I tried to sound out views on NOCs' advantages in investing abroad in two interviews in Kuwait and Saudi Arabia:

> VM: "Do you think that foreign investment by NOCs is more politically acceptable in host countries than investment by IOCs?"
>
> Kuwaiti: "I'm not going with that. Making profits for the host countries is what matters. Otherwise, it looks like we're wasting their natural resources, which would upset people."
>
> Saudi: "You say NOC, but I don't see us as an NOC. Internationally, we do our business commercially."

The reason for the frequent refusal to be labeled a national oil company is the NOCs' poor reputation within the oil industry, and even within their own organizations. They are perceived as rent absorbing, bureaucratic and politically motivated. In view of this, it is hardly surprising that the current trend among Middle Eastern NOCs is one of emancipation from the NOC model. There is a slow transition toward introducing more commercially driven practices, though the companies' ambitions in this direction fall well short of the IOC model. Some companies are rationalizing their operations, letting go of costly non-core activities; others are reducing their number of employees. But as we saw earlier, management is not prepared to introduce serious cost-cutting measures across the board. After all, it has little legitimacy to do this, being the "national" oil company.

Furthermore, for all their wish to be seen as independent commercial entities, closer inspection reveals that these NOCs do play on their government's relationship with the host country's authorities to obtain deals. For instance, the Kuwaiti government has supported KPC in its negotiation with GCC member states for the import of gas. More generally, Kuwait has had a tradition of lending money, and this has built trust. The Iranian government has given similar support to NIOC, with direct government-to-government appeals to support its activities abroad.

With regard to international upstream investments, recognized technical and managerial competence on the part of the investor is key to winning a bid to operate a field. But contrary to the NOCs' wish to be regarded primarily as commercial entities, support from the home country does help. Sonatrach has ambitiously embarked on an internationalization strategy and feels confident that it stands to gain from being the Algerian national oil company. Its employees seemed to know what host countries want to hear: "The advantage of an NOC is that we make others more serene about the fact that we will listen to their needs and understand them. There is a cultural proximity between NOCs and host countries. Also that we will not be too greedy (*gourmands*)—[they recognize] this is our reputation." Nevertheless, Sonatrach executives are aware that NOCs have to prove themselves to be as competent and efficient as IOCs. As one Algerian professional explained, "There is a perception of risk in doing business with NOCs that is not the same with IOCs (in particular with super-majors)." Sonatrach needs a successful outcome in its international ventures in order to counter this perception and increase its credibility among its peers.

For Kuwait Petroleum International (KPI), which operates in Europe, its association with Kuwait and its NOC status were obstacles to its expansion strategy. There was concern among competitors (and European governments) that the company would be supplied with Kuwaiti crude below market prices as a way to subsidize its refining. However, KPI has apparently kept suppliers at arm's length. Another disadvantage in Europe was the company's affiliation with a Muslim, Arab country, which was perceived as a potential hindrance to increasing its market share. As a result, it rebranded there as Q8. The company's studies showed that the Q8 image is of a youthful, vibrant brand, not a state brand. It wants to keep it that way.

Making the transition to an efficient commercial company while maintaining a focus on national responsibilities is the great challenge that national oil companies face today. They want to emulate the IOCs, but they do not adopt their practices. For instance, firing employees, imposing strenuous work hours and instilling a competitive (almost combative) work environment are common in the highly competitive private oil and gas industry. NOCs want to rank among the industry leaders, but are neither willing to adopt IOC standards nor ready to capitalize on their specific assets (which may be related to their relationship to the state) in order to compete with the IOCs as NOCs. However, this attitude is likely to be short-lived as NOCs observe the success of their peers.

Perhaps the NOCs' greatest asset is their unique long-term perspective. Because they have assured access to their countries' reserves and are not thinking of the next quarter, they have the luxury to think strategically, and they have the time to implement their strategy. Saudi Aramco is most deliberate in capitalizing on this asset. The company gave a striking example of its application: having established the strategic goal of expanding its downstream activities in Northeast Asia, it sent young Saudis for training and education to China, South Korea and Japan, to learn the language and the culture. To penetrate these markets, the company would need new leaders who "understand the cultures abroad." On that level, it appears that Saudi Aramco was successful: a Petronas manager told me of her surprise and admiration when receiving a visit from a Saudi Aramco counterpart who spoke to her in fluent Mandarin (her native tongue); he was calling on various producers in order to discuss developments in the regional market.

Their long-term view also gives the NOCs a different perspective on their investments' discount rates, that is, the rate at which capital depreciates over time. Private companies which operate a field for 30 years will have greater discount rates, and they will not want to leave resources in the ground for too long. In an NOC, an engineer's outlook regarding the development of reservoirs is guided by a sense of responsibility for future generations; and from an economist's perspective, the value of barrels of oil left in the ground does not decline at the same rate for NOCs as it does for private companies. As a senior director pointed out, this has important implications for reservoir management, and he felt that Saudi Aramco used "velvet gloves" in managing its reservoirs.

Investments in other areas may also follow a "national rationale" and be guided by a preoccupation for the welfare of society and state. If a company is certain to operate within a national environment for a very long time (that is, for as long as reserves are available to meet demand), it will pay greater attention to promoting the stability of that environment. Some managers felt that this explained why even commercially minded NOCs spend on the "national mission" on their own initiative. For instance, it is in the company's interest to build up a qualified pool of national candidates for recruitment. According to quite a few, the challenge is to explain to the political authorities, by nature short-term in outlook, that the results, although not immediate, are nonetheless worthwhile.

In foreign ventures especially, the long-term perspective of NOCs may set them apart from IOCs. As a Saudi Aramco employee involved in negotiations to invest downstream in China explained, the company is a patient investor. In addition to the lengthy approval process (10 years and counting), they were asked to employ a large number of Chinese to work in and around the refinery. "At this point, IOCs are not interested. NOCs are better suited for that."

Other key NOC assets are really the government's assets, its access to reserves. Sonatrach's management was very clear that the company's strengths were closely tied to its status as an NOC and to the country it belonged to. It also enjoys the strategic advantage of Algeria's geographic location, at once European, Mediterranean, African and Arab. Perhaps only in Algeria did the oil professionals feel that their company's status as a national company was an advantage, especially in its international operations. Iran's NOC benefits from the country's strategic location, which would allow the development of gas pipeline networks to India through Pakistan and to the Mediterranean through Turkey.

Other assets of the Middle Eastern NOCs have developed from responding to the challenges of the national environment in which they were created. For instance, Kuwait is a small market, and this has led KPC to focus on international markets and to hone its marketing skills. Saudi Aramco's clear asset, which often came out in discussions, is its people and their loyalty to the company. ADNOC has always worked closely with its foreign partners, and has developed superior management processes. Despite a perception in many international oil companies that most NOCs are technically incompetent, they have in fact honed specific skills that relate to the geological characteristics of their reservoirs. Iran, for instance, has a concentration of carbonate reservoirs. The constraints of these unusual reservoirs have allowed NIOC to develop specific expertise, which a senior executive felt the company could apply elsewhere in the Middle East and the Caspian Sea area. Algeria has had long experience of exploring for oil under geologically challenging salt domes. Table 3-2 recapitulates these identified company-specific strengths

Table 3-2. National Oil Company Strengths

Strengths	*ADNOC*	*KPC*	*NIOC*	*Saudi Aramco*	*Sonatrach*
Reserves	High ratio of oil and gas reserves to production and population	High ratio of oil reserves to production and population	Very large oil and gas reserves	Very large oil reserves, large gas reserves and multiple grades of crude	Oil and gas reserves, high-quality grade of crude oil
Special skills	Cooperative relations with foreign partners	International marketing skills	Negotiation skills	Long-term strategic view, human resource development	Capitalization of NOC status abroad, human resource development
Technical skills	Investment in technology and R&D	Experience with refining of sour crude oil	Experience with carbonate reservoirs	Prudent reservoir management, investment in technology and R&D	LNG expertise, experience with salt domes
Geography	Proximity to Qatar's gas fields	Access to Persian Gulf and land access to Iraq	Access to Persian Gulf and Caspian Sea, land access to Turkey and Pakistan	Access to Red Sea and Persian Gulf	Access to Mediterranean Sea
Organization	Consultative management processes	Efficient downstream business	Easy communication between subsidiaries and with Ministry of Petroleum	Efficient	Transparent accounting

and includes other assets that I observed during my field work (but it is not exhaustive).

Conclusion

The five national oil companies have various assets on which they can, and increasingly do, capitalize to become first-class oil companies. Their new commercial ambitions promise to affect their corporate culture and management processes in ways that have not yet been fully understood. Tensions are appearing between their existing business style, which is oriented toward national welfare, and is relatively comfortable (allowing women to take an important role in the companies), and the perceived need to control costs and adopt private business practices in order to fulfil their ambitions. The following three chapters examine how the national setting can support or hinder their ambitions.

four
Who Is Driving This Train?

Producing countries seek what often seems to be an elusive balance between government and national oil company. The former needs control; the latter needs autonomy. Governments control the NOC through policymaking, which includes setting targets and industry rules. They must also develop institutions able to hold the NOC accountable for its performance. National oil companies want sufficient autonomy to devise a strategy and conduct their operations. They need a clear mandate and unobstructed management processes in order to deal with industry challenges effectively and to rise to the challenge of competition. Difficulties arise in areas where decisionmaking responsibilities are shared between government and national oil company. Producing states must be clear about which bodies determine oil and gas policy and strategy, set oil prices, award licenses to international oil companies and audit the performance of the NOC and of IOCs where they are active. The clearer the limits of each institution's responsibility, the more harmonious the relationship between them will be. This chapter examines where and how national oil industry decisions are made, how the industry is held to account for the execution of those decisions and prospects for reforms of state–NOC relations in our five case studies.

Making Decisions

The Difference between Policy and Strategy

Good governance of the national petroleum sector demands a clear distinction between the roles of each institution. In this context, policy for the petroleum sector is a government prerogative. It includes setting production targets, health, safety and environment (HSE) standards and social investment targets (for example, national employment). Strategy is the plan of action by which the operator, that is, the national oil company, the international oil company, or both set out how they will achieve the targets established by government. The potential blurring between the roles of government and NOC arises because the state is the shareholder of the company and, as such, participates in the strategy-making process. The state may indeed be represented on the supreme petroleum council (SPC), which approves the strategic plan, and on the company's board, which manages day-to-day operations. If the state is involved excessively in the management of operations, the national oil company's decisions will be relatively more influenced by political objectives, presumably to the detriment of commercial considerations.

I asked both NOC managers and oil ministry officials in the five producing countries under study to explain the difference between strategy and policy, expecting the answers to put policy in the domain of the government and strategy in that of the NOC. But I found that distinguishing between policy and strategy is a classic problem and that almost all respondents struggled to explain the difference. This is partly because of the state's official, shareholder role in the NOC's strategy making, as mentioned, and the informal connections between people in both institutions. It is also because, with the exception of Saudi Arabia and Abu Dhabi, the top leadership had not made clear to most people interviewed in the ministries and the NOCs what the role of each institution was. A sample of answers to this question illustrates the confusion:

> Algerian ministry: "Maybe you could say that strategy is to ensure the short-term production of hydrocarbons and that oil policy is the hydrocarbons policy in the context of other producing countries. . . . But really, they are so intertwined. It is impossible to say where one stops and the other starts."

> Iranian NOC: "Policy is the external activity and it's done by the ministry—taxes, for example. We usually deal with the daily policy. Policy is usually internal and strategy is external."

Those who understood the difference between policy and strategy in principle often felt that reality was far more complex.

> Algerian NOC: "Oil policy is for the legislature and strategy is for the actors, the practitioners. In reality, there are no borders, no walls between them. The ministry participates in strategy because it represents the state as the shareholder of the company and also because the ministry sets out the rules of the game through policy. Sonatrach contributes also to policy because it has the expertise, and it's in the field."

> Algerian NOC: "There is no difference. Especially now, when the NOC is deeply influenced by the state (*imprégné par l'Etat*). It has no independence."

In Kuwait, the view was often expressed that strategy should be the domain of KPC, but it was unclear who should handle policymaking.

> Kuwaiti NOC: "In 1980, when they created KPC and the SPC [Supreme Petroleum Council], it weakened the role of the ministry. The SPC is the general assembly of KPC. It gives the strategic policy guidelines. So the ministry is left wondering 'what is our role?'"

In Saudi Arabia and the United Arab Emirates, the difference between policy and strategy was clearer than elsewhere because the respective prerogatives of the oil ministry and the NOC are clear. In Abu Dhabi, oil strategy is "a local issue"; it is not a federal (UAE) issue, and its management stays in Abu Dhabi. As a senior ADNOC executive explained, "The SPC sets strategy for the local level. The ministry is at the federal level and coordinates policy." Policy issues are limited to OPEC representation and subsidies at the pump. With similar clarity, Saudis explained that the oil ministry is responsible for dealing with OPEC, supply and demand, opening the hydrocarbon (currently non-associated gas) sector to private investment and the consumer–producer dialogue. A vice-president at Aramco explained that "'strategic imperatives' are . . . within our control, like improving our performance, optimizing our portfolio, capturing markets, protecting markets, expanding the local economy, preparing the workforce for the future."

Policymaking Processes

Policy directives from government normally pertain to oil production volumes, environmental laws, targets for employment and the promotion of the private sector, and oil conservation guidelines. In addition, companies may receive general directives on maintaining strategic market share. This is the case in Saudi Arabia. Several Aramco executives explained that the company maintains its share of the less profitable European and American markets in order "to keep relationships there." The government may also ask it to increase its market share. In these instances, Saudi Aramco is not pursuing a bottom-line strategy, but it is playing a crucial role in supporting the state's priorities.

In shaping policy directives, the oil ministry will often request feedback from the NOC on the substance of directives that affect the oil industry. In Algeria, for example, when a law is drafted, the Ministry of Energy and Mines sends it to Sonatrach for review, usually approaching the relevant managers for comments and critiques. Behind the formal process of decisionmaking, the company's corporate planning department works closely with the ministry on policy planning in most cases. The managers involved see this cooperation as positive. In other countries too most company managers and ministry officials spoke of collaboration and information exchange as being for the benefit of a better oil policy. On an informal level, there are numerous ties between the ministries and the NOCs, as many of the ministry people, often including the minister himself, have come from the national oil company.

In Saudi Arabia, as in Iran and Kuwait, the government has also asked the NOC to provide advisers for the ministry of energy and to take part in the government's strategic committees. These advisers help to bridge the knowledge gap between the ministry and the company. Nevertheless, it was clear from interviews that some ministry officials felt unable to move the imposing machine of the NOC.

With the exception of the Emirates, where the oil ministry plays a limited role, collaboration is particularly prevalent in deciding whether or not to raise production capacity, as the ministry will ask the NOC to give indications of requirements and capabilities for increasing capacity. Sonatrach, for instance, explained that it would first make a proposal, based on its data, which both bodies would then discuss. A corporate planner explained that the relationship between Sonatrach and the ministry was friendly and easygoing in such matters. In Saudi Arabia, oil minister Ali Naimi said that on such decisions, there is dialogue with Saudi Aramco but that ultimately these decisions are led by government policy, which aims "to meet international oil demand and stabilize the markets."[1]

1. Comments made by the Minister of Petroleum and Mineral Resources, Ali Naimi, at the conference "Oil, Economic Change and the Business Sector in the Middle East," Chatham House, London, 29–30 November 2004.

In the case of the Emirates, ADNOC prepares plans with various options for raising capacity. These are informed by a lengthy and careful consultative process within the operating companies and subsequently the parent company. Then, it presents the plans to the SPC on demand. It was apparent from interviews that the decision to raise capacity was informed or driven not by OPEC but by the initiative of the leadership—that is, the ruler. These plans for capacity increases are usually broad policies that the Supreme Petroleum Council translates into strategic guidelines.

In Kuwait, decisions about raising capacity appear to go back and forth between institutions, with a degree of confusion for all concerned. This underlines the tension between institutions. Upstream and corporate planning managers at KOC and KPC explained the process by which they arrived at their present production capacity targets: KPC's corporate planning department determined an initial range for maximum expansion of production capacity; this was based on its analysis of global supply and demand projections. The analysis showed that the call on Kuwaiti oil would reach 7 million b/d by 2020, and this led KPC to ask its operating company whether this target was feasible. The Kuwait Oil Company responded that 4 million b/d was the maximum production level it could sustain. On the basis of this information KPC presented a strategic plan to deliver 4 million b/d to the SPC, which decided that the company should aim for 5 million b/d. Within the company, this higher target was not seen as sustainable. The SPC eventually reverted to the 4 million b/d target.

Though there have been some institutional difficulties in Kuwait, collaboration on matters such as raising capacity generally appears to be easy, especially in contrast to granting international oil companies licenses to develop petroleum resources. Licensing is an area of responsibility that has been fought over in several countries. In the course of negotiations for new concessions in Algeria and Saudi Arabia friction arose between Sonatrach and the oil ministry and between Saudi Aramco and some government figures (see "Reforms" below).

Usually, the NOC is not consulted on key oil policy decisions—that is, on decisions relating to OPEC politics and policy. When asked about specific policy points, Saudi Aramco managers tended to refer to the ministry: "We don't set government policy (in relation to OPEC in particular). We make sure we don't get sucked into their process. It's better to divide these roles. We deliver the goods. The government can manage the relationships and keep the supply and demand balanced." In fact, throughout the region, key policy decisions appear to be not in the hands of the ministries of petroleum but in those of the rulers, as the following comment from a senior Kuwaiti OPEC representative indicates: "OPEC is the government's responsibility but the decision is taken at the highest level."

On another front, decisionmaking authority for the pricing of crude oil exports, which had previously been the exclusive domain of the ministries, has gradually shifted to the NOCs.[2] A high-ranking ministry official in Saudi Arabia explained that the responsibility for export crude pricing has been completely in Saudi Aramco's hands since 1986. An Aramco marketing executive added that crude pricing is now a routine decision and that a committee meets to determine the "quality adjustment" (a premium or a discount). The authority to determine the price formula lies with the NOC in Algeria and Iran as well. In Iran, the right to approve this formula is vested by law in the minister, but he has given it to the managing director, who entrusts the international affairs department with the decision. As for Abu Dhabi, ADNOC's crude oil entitlement is sold on the basis of a retroactive price.[3] Pricing is therefore quite a straightforward commercial process: the company makes price recommendations to the SPC on the basis of available market data.

Most company marketing teams felt that this decentralization of decisionmaking about pricing gave them the necessary flexibility to respond to the market, because it minimized delay and kept decisions commercial. But sometimes, the commerciality of pricing methodologies has to be defended by the NOCs. A Saudi Aramco manager, for example, explained that the company has had to fight against a domestic price policy based primarily on the needs of the government.

Only in Kuwait does the oil ministry tell the company what the price of a barrel of oil should be. A price formula is devised every five years, and the ministry's pricing committee then calculates a price each month, in the same way as Saudi Aramco does. A finance director at the Kuwait Petroleum Corporation pointed out that the ministry's decision (which remains an assumption) on price per barrel could be problematic if KPC feels that the market would not support that price. "We want the assumed price to be close to the market price." Tension can arise if the assumed price is higher than the market price. Indeed, KPC sells the crude for the ministry and takes only a small commission on this sale. Therefore, it stands to gain more from a lower crude

2. The NOCs sell most of their crude oil on term contracts, though the terms are short (3–18 months is typical). The prices are derived from a "market-related" benchmark, such as the Brent or WTI commodity prices, with adjustments for timing, quality and the point of sale.

3. Retroactive pricing means that the price of a barrel is based on average market prices during the previous month.

sale price because this means a higher income is projected for its refining operations, which could purchase the crude. Kuwait may change this system in favor of giving the company revenue targets rather than price targets.

In all the case study countries, product pricing is less politically sensitive. It is handled within the NOC and is based mostly on market prices quoted by industry publications such as *Platts*. It also appears that customers have a greater say in product pricing than in crude pricing, for which decisionmaking is more centralized and there is only a little flexibility with the discount.

On lower-level issues, such as assessing the supply and demand picture, the Kuwaiti oil ministry and KPC "coordinate fully." This is specific to the Kuwaiti system, in which there is some blurring of responsibilities on OPEC policy because until recently, the OPEC governor held a key position in KPC. This is likely to change in the coming years (see "Political Interference in Operations" below).

The Strategy Decisionmaking Process

There is a formal process for strategy and operational decisionmaking through which the relevant state institutions can influence the management of the hydrocarbon sector. In most cases, the highest decisionmaking body of the NOC is the board, on which the company's shareholder (the government) is represented. Above this sits a decisionmaking body, often called the Supreme Petroleum Council, whose role is to oversee the hydrocarbon sector and to avoid deadlocks between company and government. The composition of these bodies (see tables 4-1 and 4-2) gives an indication of the measure of separation that exists between government and company and the interests that may influence or drive decisionmaking. For instance, in all countries with a traditional monarchic regime, Kuwait, Abu Dhabi and Saudi Arabia, the political leader or the crown prince chairs the Supreme Petroleum Council. This indicates a highly centralized decisionmaking structure. However, some of the formal processes do not reflect the true centers of power. For instance, in all cases the minister of petroleum chairs the board of directors, except in Iran, where that role falls on the managing director of NIOC. And yet, in reality, the ministry dominates the decisionmaking process in Iran.

Saudi Arabia's oil industry operations are overseen by Saudi Aramco's board of directors; the Supreme Petroleum Council governs the strategy decisionmaking process. A unique feature of the board is that it includes three former CEOs of international oil companies for the purpose of providing external advice. This is a legacy of the privately owned Aramco days and

Table 4-1. The Composition of the NOC Board of Directors in Abu Dhabi, Kuwait, Iran, Saudi Arabia and Algeria

Board of directors	*Government representative*	*Company and non-government representative*
ADNOC	n/a	n/a
KPC	Chair: Minister of Energy Seat: Ministry of Finance	Deputy chair: KPC's CEO Seats: seven MDs, including subsidiary heads (not KUFPEC)
NIOC	Seat: Ministry of Petroleum (OPEC representative)	Chair: NIOC's managing director Seats: Deputy MDs and six directors of operations
Saudi Aramco	Chair: Minister of Petroleum and Mineral Resources Seats: Secretary General of Supreme Economic Council, two ministers, VP of King Abdul Aziz City for Science and Technology	Seats: Saudi Aramco's CEO and three VPs, former minister of state and president of the seaport authority, retired presidents of Marathon and Texaco, former vice chairman of J.P. Morgan & Co.
Sonatrach	Chair: Minister of Energy and Mines Seats: Governor of central bank, other ministers	Seats: CEO of Sonatrach, CEO of Sonelgaz

Source: Author's interviews.

Saudi Aramco's enduring links with American companies. Saudi Arabia is also the only country in this study to have representatives of educational institutions on its top hydrocarbon committees.

As for the Emirates' governance of the petroleum sector, the old Department of Petroleum and the ADNOC board were dissolved after a presidential decree in 1988. The functions of these bodies (the administration and supervision of the country's petroleum affairs) were taken over by the newly formed Supreme Petroleum Council. ADNOC's 18 operating companies each have boards (which include foreign shareholders), and they are responsible for budget and business plans and the implementation of strategy decisions from the SPC. The heads of ADNOC's eight business directorates report directly to the CEO, but there is also a petroleum steering committee, which may assume some of the responsibilities of a board of directors.

The Algerian oil and gas sector's organization is somewhat different. The National Energy Council (NEC) was established in 1981 to take charge of both energy policy and strategy. But the ruling party gained a stranglehold, and its heavy influence severely restricted Sonatrach's market flexibility. The NEC was reformed several times, and in the process it lost a number of powers to the Ministry of Energy and Mines. The ministry sends representatives to the board of directors (Conseil d'administration) and to the General

Table 4-2. The Composition of the Supreme Petroleum Council in Abu Dhabi, Kuwait, Iran, Saudi Arabia and Algeria

Supreme Petroleum Council (or equivalent)	*Government representative*	*Company representative*
Abu Dhabi	Chair: President of UAE (previously the crown prince) Seats: Governor of central bank, Abu Dhabi Investment Authority representative, eight Abu Dhabi government officials	Secretary General: ADNOC's CEO Seat: ADNOC's deputy CEO
Kuwait	Chair: Prime minister Seats: Governor of central bank, five ministers	Deputy chair: KPC's CEO Seats: nine nongovernment members
Iran	Chair: Minister of Petroleum Seats: Other ministers	Seats: NIOC's managing director/ deputy minister of petroleum, NIOC's deputy MD
Saudi Arabia	Chair: the king (prime minister) Deputy chair: the crown prince (deputy prime minister) Seats: eight ministers	Seats: Saudi Aramco's CEO
Algeria (General Assembly)	Chair: Minister of Energy and Mines Seats: Governor of central bank, a representative of the presidency, three leading ministers	None

Source: Author's interviews.

Assembly, which is similar to a supreme petroleum council, but not to the executive committee. This consists only of Sonatrach officers; it manages day-to-day operations. The board of directors is concerned with corporate governance and acts as a buffer between the General Assembly and the executive committee.

In Abu Dhabi, Algeria, Kuwait and Saudi Arabia, the operating companies initiate most of their project decisions, which must be approved by the supreme petroleum councils (or their equivalent). Saudi Aramco enjoys a great degree of leeway in this process. It prepares a business plan every year covering major strategies and plans over a rolling five-year horizon. The Saudi SPC approves the strategy plan as presented by Saudi Aramco via its board of directors (the board). The strategy plan has always been approved. A senior oil ministry official commented, "It is symbolic because [the SPC] always approves the decisions made by the board; it is not reprimanded. This is because there's an unwritten aura, a taboo culture against interference [in the oil sector]."

The situation is different in Kuwait, where some KPC board members feel that the lines of communication and reporting are unclear because directives are sent from multiple sources. Indeed, the KPC board elaborates a strategy on the basis of the directives sent by both the SPC and the Ministerial Council. KPC's strategy must reflect the SPC's general directives and the Ministerial Council's policy objectives, which are conveyed by the Ministry of Energy.

In Iran, by contrast, NIOC managers explained that the company must follow government directives on what projects to pursue. As the government is responsible for NIOC's budget, it seems that there is no room for the company to negotiate on spending plans.

Political Interference in Operations

Outside these formal processes, NOCs often receive other types of government "request." In Kuwait notably, political interference hampers operations. Kuwaitis in and out of KPC refer to their members of parliament to obtain various favors. As a result, personal, political and institutional interests become intertwined, with damaging effects on the management of the oil industry. One can also draw on family networks for professional gain. Nepotism and institutional interference have caused some mid-level managerial positions in the company to be filled by inexperienced Kuwaitis. An executive commented on this political interference in management decisions: "People need to have loyalty to KPC. If an employee doesn't get the promotion he wants, he can go to the minister, to members of parliament. The family will interfere. The company promotes people who don't deserve it. We need a strong management to counter that. We need an agreement with government on the rules."

Political interference is a common problem in the capital-intensive oil industry. Saudi Aramco, by contrast, appears to be insulated from government requests for favors and special treatment. When asked how the company had succeeded in this, officials from the oil ministry and Saudi Aramco managers all pointed to the strong message against interference sent by the high leadership. A senior official explained, "We have instructions from the king: no one is to interfere with Saudi Aramco. There is to be no direct contact, and any communication must be through the ministry. And the ministry has instructions not to interfere in the day-to-day management of Saudi Aramco." The "daily business" of the company includes deciding to whom the company sells oil and setting prices and contract terms. The board supervises decisions on such matters. A senior figure at the ministry added, "We all gave Aramco an aura of protection. The higher up in the political ladder, the more they wanted the company to be independent. [This attitude is] inherited and is reinforced." Another official felt that the "good people leading the ministry," such as Ali Naimi and Hisham Nazer,[4] also contributed to this self-restraint. The general feeling in Aramco (and apparently in the ministry too) was that the relationship with government is healthy. As one person put it, "It gives both latitude to the company and the control that the government wants."

The Abu Dhabi (and UAE) political system is highly centralized and cohesive. High strategy decisions are made by government (that is to say, the ruler), and these are executed by the NOC. Because of the unity of purpose, there is no need for government "interference" in ADNOC's operations. "The president, the Abu Dhabi government and the UAE are one. Sheikh Zayed is the decisionmaker for all this. The same body decides," an executive explained.[5]

In Algeria, numerous managers in Sonatrach said that the state intervenes in everything the company does. But it appears that once the state was even more interventionist—to the degree that it is now in Iran—and had deep and far-reaching control over the network of national companies. A manager explained that monopolies then dealt with each other, which allowed government to intervene at various levels and to set the price of oil. Now, privatization is slowly gaining pace; and even though the private sector remains limited, this trend is seen as a necessary step to prepare the country for globalization. Parallel to this, the national companies are increasingly treated like any other company: Sonatrach's CEO apparently no longer has direct access to the president of the republic.

Operations: Is the NOC Doing a Good Job?

Petroleum resources are entrusted to oil companies. The state must ensure that its finite hydrocarbon reserves are developed responsibly. Damage done to a reservoir can mean that some of the oil it contains becomes irretrievable. Companies are required to comply with general oil policy, which includes HSE standards and targets for the employment of nationals. There

4. Mr. Nazer served as Minister of Petroleum from 1986 to 1995.
5. This interview was conducted in September 2004, before Sheikh Zayed's death.

must be a body that audits the performance of industry in this sector of crucial importance to the economy. This audit would apply to the national oil company and to any private investors active in the sector, such as international oil companies.

Potentially, there are a number of bodies that can carry out this role: the national oil company, the ministry of petroleum, an independent regulatory agency and parliament. In countries where the NOC is the regulator of the industry, an obvious problem of conflict of interest emerges. It cannot regulate itself, nor can it regulate foreign oil companies in activities in which it is itself a partner. The other options create more checks and balances in the system, provided technical competence and the independence of the auditing institutions is ensured.

Government Oversight

The transfer of regulatory responsibilities from NOC to ministry of petroleum (or to an auditing/regulatory agency) requires efforts to bridge the knowledge gap between it and the national oil company. Most ministries of energy have NOC experts on secondment to assist in technical matters and the transfer of skills. On the whole, however, the ministries are in need of competent people to set policy objectives and to audit operations. Though ministry posts are not as well remunerated as those in the oil companies, a Kuwaiti ministry official noted that ministry positions carry more social influence in her country. In Saudi Arabia, positions with the NOC are more coveted. The strict selection process of the best students in the kingdom for entry into Saudi Aramco's training programs harnesses the country's best minds to the oil and gas sector.

In the case of the Norwegian Petroleum Directorate (NPD), it took about 15 years before this regulatory agency was able to recruit top people from the international oil industry. But eventually, Norway was able to establish a strong regulatory body. The NPD's governance model applies the idea of democratic checks and balances to the management of this coveted sector. It gives the oil ministry the role of defining targets and setting standards through making policy. The NOCs and IOCs are confined to the operational field, which includes devising strategy. The NPD itself has the responsibility of auditing the companies' adherence to the standards set by government. Even so, it looks to the national oil company, Statoil, for "objective," loyal advice, and the NOC continues to carry national responsibility for the resource. Statoil would, for instance, have special responsibility to make the government aware of market and technical issues, so that the latter can make

Table 4-3. Regulatory Bodies in Major Hydrocarbon-Producing Countries, 2004[a]

Regulatory function/ Regulatory body	*Licensing*	*Oversees NOC budget*	*Oversees development plans*	*Regulates upstream activities*	*HSE*
Algeria: ALNAFT	•		•	•	
ARH					•
Brazil: ANP	•	•		•	
Bolivia: SH	•		•		•
Canada: AEUB				•	•
Colombia: ANH	•	•	•	•	
India: DGH		•	•	•	
Indonesia: BP Migas	•			•	•
Mexico: CRE	•			•	
Nigeria: DPR	•			•	•
Norway: NPD[b]	•		•	•	•
United Kingdom: DTI	•			•	
HSE				•	•
United States (e.g., Alaska AOGCC)	•		•	•	•

Sources: EIA country analysis briefs; Embassy of Colombia, Washington, D.C.; U.S. embassy in Jakarta; *Gas Matters.* The following websites were consulted: Alaska Department of Natural Resources; Alaska Oil and Gas Conservation Commission; Alberta Energy Utilities Board; Agência Nacional do Petróleo (Brazil); Algerian Ministry of Energy and Mines; Comision Reguladora de Energia (Mexico); Directorate General of Hydrocarbons (India); U.K. Department of Trade and Industry (Licensing Consents Unit); U.K. Health and Safety Executive; Nigerian Department of Petroleum Resources; the Norwegian Petroleum Directorate; and La Superintendencia de Hidrocarburos (Bolivia).

a. The functions identified here are derived from the main available sources (for which see above). Regulators may have other functions that were not expressly mentioned on their or their ministry's official website.

b. In January 2004 the NPD was split to form the Petroleum Safety Authority (PSA) Norway, which reports to the Norwegian Ministry of Labor and Government Administration. There is also a Pollution Control Authority, which cooperates with the NPD and the PSA on environmental regulation.

informed decisions about depletion and price. This regulatory model has inspired a number of states, including Algeria, as table 4-3 illustrates.

Generally, discussions with me about the regulatory model for governance of the oil sector highlighted the need for much greater clarity in the respective prerogatives of ministry and NOC. Few, however, were in favor of a separate regulatory agency. In most of the Middle East, regulation of the industry is carried out by either the ministry of petroleum or the NOC. These organizations have, in many cases, struggled with each other to take over or to retain this function. Historically, the role has fallen on the national oil companies, but it is increasingly being transferred to the ministries, as in the cases of Saudi Arabia and Kuwait. As pointed out in the section on "Reform" below, these transfers of regulatory powers have been at the heart of reforms of the petroleum sector in most of the five countries considered in this study.

Parliamentary Oversight

In Kuwait, Iran and Algeria, the parliament and forces outside the energy sector play a supervisory or legislative role. The Kuwaiti parliament, for one, has sought to influence the government's oil policy and the operators' management of the oil sector for decades—we saw in chapter 1 the role it played in 1963 when it launched a public awareness campaign criticizing the foreign-owned KOC. In recent years, it has been concerned about both insufficient returns and accidents that have occurred in Kuwait's refineries. As a result, members of parliament have put additional pressure on the government to regulate the industry. This pressure is generally seen as positive, to the extent that it has brought about an improvement in HSE practices in the NOC. However, it has also meant that decisionmaking processes have slowed down significantly, as parliamentary committees question a number of company decisions. All of KPC's investment plans and draft contracts have to be submitted to the National Assembly, after they have been approved by the KPC board, the oil ministry and the SPC. Feelings about this are mixed.

> KOC: "The parliament is sometimes good for us. Sometimes we wish there wasn't a parliament."
>
> VM: "When?"
>
> KOC: "They're good for us because they bring more transparency (in tendering, for example); they control abuses of power. They're a watchdog. But sometimes they misuse democracy. They want attention or deals, personal favors."

The personal politics of Kuwait's political system are compounded by the weakness of its political parties. Furthermore, the government does not have a majority in parliament. It must, as a result, negotiate for the support of members of parliament on every issue, causing those outside the cabinet to become more influential than those inside. In Algeria, by contrast, the National Assembly's debate on the reform of the hydrocarbon sector was structured by the presence of well-established political parties.

Also, it is questionable whether the Kuwaiti National Assembly has the technical competence to regulate the oil industry. The following comments made by employees of KPC and its subsidiaries indicate a degree of frustration. "It would be better to give us a percentage of the profits, to get agreement on the projects and the targets and then to audit us two years later—not to be step by step with me!" "The strategy is approved on a general level. [The investment decision] shouldn't [need to] be reviewed by the SPC and the parliament in detail! Our board of directors has to take back its powers and its independence. . . . Policymakers shouldn't reverse their decisions. It leads to confusion."

In Iran, the Majlis has until now looked very closely at the issue of gasoline prices. There is a great deal of uncertainty about the new (seventh) parliament and the position it will take on the bill proposing a reform of the subsidy system for gasoline[6] and on a broader plan to separate the Ministry of Petroleum and NIOC (discussed in greater detail in the next section). The challenges to these reforms, as I gathered from interviews, did not come from one interest group in particular. The following exchange helps to describe the situation.

VM: "Who opposed this reform?"

IR10: "Parliament."

VM: "Who in [the fifth] parliament? Which group?"

IR10: "Everybody!! They wanted to be re-elected. They were not acting in the national interest."

VM: "But the sixth parliament approved the reform . . . Why was that?"

IR10: "Elections for the seventh parliament were in December–March and the discussions on this bill were in April–May–June. So everyone there knew if they were staying or leaving. So they took a decision in the national interest!"

IR11: "In parliament, people think the resource belongs to the people. They are not thinking of the economy. They think about nationalism and populism."

IR10: "Actually, it's a disaster for our country. We need capital formation."

IR11: "The low prices increase [gasoline] consumption."

IR10: "Our consumption is higher than anyone in the world (by share of GDP)."

6. See John Mitchell's special contribution on the economic background at the end of this volume for an analysis of the costs of subsidies to the Iranian economy.

Uncertainty about the proposed reform is compounded by the election of Mahmoud Ahmadinejad to the presidency. It is unclear how the conservative parliament will interact with a conservative president. From events unfolding in late 2005, it appeared that the parliament was prepared to counter some of the president's decisions regarding the oil and gas sector (for example, his first three candidates for the ministerial post were rejected because they were deemed insufficiently qualified).

The Iranian and Kuwaiti parliaments have also been critical of foreign investment in their countries' hydrocarbon sector. The pressure is uncomfortable for those in the NOCs who negotiate with the foreign companies, as an Iranian explained: "Here we have a parliament that looks at me like a traitor for negotiating foreign participation in the oil industry." Iranian MPs are currently studying a new bill that would give the Majlis more supervisory powers over buybacks and other types of foreign investment contracts. Moreover, senior legislators already press the national oil and gas industry, by way of the Majlis Energy Commission, to cut costs and improve efficiency. The commission is also an anti-corruption watchdog.

The media in Iran are also keen observers of the energy scene. Daily papers have regular columns and full-page sections on energy news. They question decisions made by NIOC and the energy ministry and have followed the debate on gasoline subsidies closely. One manager mentioned that the papers question NIOC's salaries: "They say, 'Why should the people working at the ministry of energy or in NIOC get a salary three times higher than in the Ministry of Agriculture?'"

In Kuwait, the Supreme Petroleum Council devised Project Kuwait in 1997 as a way to raise production by inviting foreign companies back to Kuwait in developing its northern oil fields. But before bids can be invited, parliament must approve an enabling law. A first draft law was rejected in 2002. Influential members of parliament opposed to foreign investment in Kuwait's upstream have argued that the project goes against the constitution's decree on public ownership of natural resources. A key concern raised in parliamentary debates has been the prospect of the loss of jobs for Kuwaitis. Most members of parliament agreed that the introduction of foreign private investment would force a rationalization of activities and an elimination of what is delicately called "disguised unemployment" (Tétreault, 2003: 89). Maintaining local content in the industry's outsourcing was another main subject of the debate.

In 2003, the Kuwaiti government revived a higher committee to work on the revised draft of a foreign investment bill in which the fiscal terms and a draft contract for IOCs were outlined. This committee was formed under the oil minister, with KPC's deputy CEO as vice chairman. It included the oil ministry undersecretary and the chairmen of the main upstream companies in Kuwait, including Ahmad Al-Arbeed (formerly of KOC, now leading Project Kuwait). The revised draft was recently submitted to the SPC, and will be reviewed by parliament. Clearly, the need to gain parliamentary approval has influenced the contractual terms. The new draft document attempts to satisfy public concern about the local benefits of foreign investment by setting specific targets for local content in the contract terms. The energy minister, Sheikh Ahmad Al-Fahd Al-Sabah, explained that the terms and conditions of Project Kuwait will require that 60 percent of the initial workforce employed in developing the northern oil fields must be Kuwaiti citizens and that this requirement will increase in time to 80 percent. Moreover, the draft law obliges the winning consortium to include a Kuwaiti state-owned entity. It restricts foreign investment strictly to the northern oil fields, ensuring that the Burgan field, Kuwait's "crown jewel," remains the preserve of the NOC. The draft states that the foreign oil companies will work under an operating service agreement, with cash reimbursement rather than payment in oil.

Reforms

Algeria

At the time of interviewing in Algeria, in January 2004, the presidential elections were four months away and the reformers' position was uncertain. The hydrocarbon reform, led by the minister of energy and former World Bank economist Chakib Khelil, would curtail Sonatrach's dominant position in the oil and gas sector. The reform bill had been frozen by opposition from within Sonatrach (essentially from the union, UGTA, l'Union Générale des Travailleurs Algériens) and civil society groups, concerned about layoffs and the possible privatization of the company. However, some Algerians felt that the state was too pervasive and that Sonatrach lacked independence. Supporters of the proposed reform of the sector hoped that reform would relieve Sonatrach of its ambiguous role vis-à-vis both the state and the foreign companies. But others worried about how Sonatrach would compete without its historical special rights over the mineral domain in Algeria and its authority over licensing.

President Bouteflika has since won a strong mandate, which gave the reform bill a second life. It was passed by the upper house of parliament in

Figure 4-1. Roles in the National Petroleum Sector—Checks and Balances

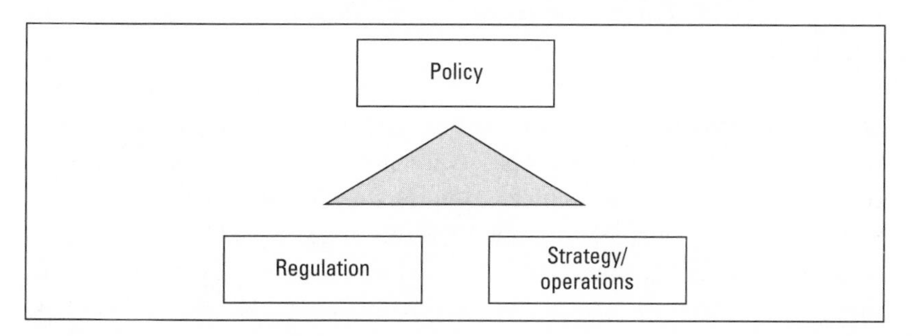

March 2005. Subsequent conversations with Sonatrach managers have indicated that this reform is now accepted by UGTA as well as by society generally. The reform gives Sonatrach an option to take a 20–30 percent stake in discoveries, down from 51 percent. It ends Sonatrach's monopoly over pipelines and the downstream business and takes regulatory responsibilities away from it.

In interviews, I drew a triangle (see figure 4-1), which is often used to illustrate an optimal demarcation of roles in the national petroleum sector. It also conveys the idea of checks and balances.

Those interviewed in Algeria recognized the triangular model as the one presented in the hydrocarbon reform bill. Their degree of familiarity with the regulatory model was unique in the region—a sign that the extensive efforts to explain it had been successful. The energy ministry and Sonatrach's management had explained the modalities of the reform to Sonatrach employees, to UGTA and to parliament over three years, having held a number of seminars and meetings, including a Q&A session with the energy minister. Interestingly, though this bill faced intense opposition from Sonatrach, UGTA and society at large when it was first presented, my interviews in Sonatrach in 2004 indicated that once it was put on the shelf a number of people across the political and industry spectrum started to accept some of its ideas and to view its future application in one form or another as inevitable. They were much better informed about the objectives and characteristics of the reform than when it was first presented; as the political storm that surrounded its first introduction gave way to a more practical environment, these facts could be more carefully considered and consensus began to grow in support of it. Interviews with several observers outside Sonatrach suggested that those of its executives who had opposed the reform were transferred out of the company by the minister.

In doing so, these observers argued, he acted to bring the executive circle more firmly under the government's control and thus to pave the way for the reform.

A major aspect of the reform is the creation of two regulatory agencies. One is the more technical Autorité de régulation des hydrocarbures. This national authority will audit the petroleum industry's health, safety and environment standards and regulate compliance with technical standards in upstream and downstream activities. The other agency, ALNAFT, will handle the promotion of exploration in Algeria, which Sonatrach has been doing for the state. Its functions will also include maintaining a large upstream database, managing the system of licensing rounds for upstream investments, signing contracts with foreign oil companies for exploration and production, and approving development plans. It will follow up and control, as a contracting party, implementation of the exploration and exploitation agreements in accordance with the law.

Within the energy ministry, I found there was support for the introduction of both regulators, as the following comments show. "My belief is that a commercial entity must take care of its own business—commerce. To each his own field. A commercial company will always act to make money and will think like a company. It can't supervise other companies. There is a conflict of interest there. The state has to be the one doing the policy. That's its business." "[Having a regulator for HSE] would have been very helpful [to prevent what] just happened in Skikda.[7] The state needs an agency that is independent and with means of control because otherwise the NOC does it and is in a position of conflict of interest because it operates the plants."

Within Sonatrach, most supported the idea of an HSE regulator. There is in fact support for this form of regulation in all the five national oil companies considered in this study because it is inoffensive and would not threaten their established interests. In Iran, for example, a planning ministry official argued in favor of a regulator for domestic energy consumption; it would be responsible as well for shaping the mix of fuels consumed in Iran in line with policy objectives and for regulating the price of energy.

But, as intimated above, a number of Sonatrach employees had initially resisted the creation of ALNAFT because it would reduce significantly the

7. On 19 January 2004, a steam boiler at a Sonatrach LNG production plant released gas from its pressure valve, which then ignited, triggering a second, more serious vapor-cloud explosion and fire that took eight hours to extinguish. Around 30 people were killed and damage to the plant ran to U.S.$800 million.

company's influence over the hydrocarbon industry—essentially its control over licensing IOC contracts, approving development plans and supervising their execution. Many of its executives felt they had the expertise and capacity to exercise the role of an independent body that selected tenders and awarded new licenses. I heard similar views in Saudi Aramco. Sonatrach and Saudi Aramco have both been criticized for taking on this role, and it has created friction with their respective ministries of energy or other government groups (for the case of Saudi Aramco, see below). Both companies have claimed that they have engaged foreign companies in a transparent and efficient bidding process and that a new body is unnecessary.[8] In fact, until the reform was ratified, the Sonatrach licensing commission selected the companies on technical merit in a public setting with media present. The transparent bidding process that was put in place before the introduction of the new law was undoubtedly a result of ministry pressure on Sonatrach and part of its attempt to ward off the perceived need for a new licensing agency.

The following comment on the process was made by a Sonatrach executive: "Sonatrach is criticized for being the operator, the selector of permits and the regulator. But really, [it does its job]. Sonatrach talks to the state to emphasize the commercial requisites. And it turns around to the IOCs to emphasize the national imperatives."

Managers in other companies shared this sense of skepticism about a new regulatory body and doubted that it could surpass the national oil company's capacity to keep foreign companies in check and to supervise operations. This supervisory function is the principal concern of NOCs throughout the region: as we saw in chapter 2, they do not trust foreign oil companies to manage reservoirs with caution so as to maximize the ultimate recovery of resources over the long term. Moreover, many NOC managers do not trust their government's technical competence to supervise this activity, nor are they confident that their government would be as careful as they are with the reservoirs (this is not the case in Iran).

Some executives at Sonatrach also opposed the creation of ALNAFT because they perceived a danger of a domestic brain drain. As mentioned earlier, the challenge of attracting the best talent is a significant consideration in setting up a regulatory agency. It must be able to recruit and pay the salaries

8. In Saudi Arabia the bidding process is not as transparent as in Algeria. Aramco evaluates the technical and commercial bids for contracts behind closed doors. The company says this is the preference of the bidding companies: "The companies wouldn't like to have their offer exposed." In any case, as we will see, some bidding and supervisory responsibilities have of late been transferred to the ministry.

of very skilled professionals. Minister Khelil's plans for ALNAFT included provisions for high salaries and other exemptions that would attract those professionals. The fear was that this would draw them away from Sonatrach and cripple it. The company would be left with a larger proportion of unproductive employees, which management would be loath to fire for reasons related to its corporate culture and popular Algerian expectations.

Saudi Arabia

In Saudi Arabia, both Ministry of Petroleum and Mineral Resources officials and company managers affirm that the boundaries between them are clear and that there is no "trespassing." It is interesting to note, however, that ministry people identify themselves as Aramcons: for instance, they often say "we at Aramco." When I mentioned this to a Saudi Aramco executive, he was quite surprised.

But institutional tensions did arise during the Strategic Gas Initiative (SGI), a program for foreign investment in the Saudi upstream gas sector launched in 1998. This tension was not so much between the NOC and the ministry as between those two institutions and other government groups hostile to the independence and strength of Aramco. For this reason, the Strategic Gas Initiative deserves specific attention as an important event in the history of Saudi energy and politics because it shed light on the distribution of power between the relevant institutions that shape the kingdom's decisions on energy matters. It also made clearer to those involved the limits of each institution's role and responsibility.

The international oil companies had long dreamt of being asked to return to the kingdom to develop its resources. It appears from discussions with Saudis that the initiative for the SGI had been political, as the then crown prince had contacted U.S. companies directly while in Washington, D.C., and made a general offer. The invitation was to develop gas, but some IOCs hoped there was more to come. It appears that the crown prince wanted to introduce competition into the sector. A Saudi Aramco executive acknowledged that the company had let oil refining and gas stagnate for a long time. "We were still coming into our own. At the end of the 80s–early 90s, there had been little investment in refining and gas. There was a frustration in the public sector (and in government generally) that we weren't showing enough in investment terms. There was little or no FDI [foreign direct investment]. Saudi Aramco was behind in producing non-associated gas. Now we have risen to the challenge. Between 1986 and 1996, we found gas and built gas plants in Hawiya and Haradh."

Though there was a great deal of pride among Aramcons for what they accomplished in that decade, most Aramco managers denied any resentment of the political initiative to invite foreigners into their preserve. They explained that they complied with the decision to encourage FDI and thought only about how to implement it. But a Saudi Aramco executive conceded that some in the company took the crown prince's offer to the IOCs as an insult, interpreting it as a lack of confidence in the company's capacity to develop the country's gas resources alone. (A foreign oil company executive involved in the negotiations saw this initiative as part of a broader plan to reform the kingdom's economy and stimulate investment at a time of low oil prices.)

The ensuing process of defining the project and negotiating with the foreign companies involved both the NOC and the government. The crown prince's initiative was picked up by a committee headed by the influential foreign minister, Saud Al-Faysal; it consisted of three representatives from Saudi Aramco and three from the ministries that would be affected by the foreign investment. Things then got complicated as the committee discussed the practicalities of implementing Prince Abdullah's decision to invite foreign investment. An Aramco executive involved in the negotiation process explained that though the foreign minister wanted to engage pre-selected companies in negotiations, Saudi Aramco preferred a one-stop bid, with terms clearly set out from the beginning so that the companies would not pursue discussions with false hopes regarding the outcome.

The committee also discussed Saudi investment needs and focused on gas development, desalination, petrochemicals, refining and power generation. Gas investments were part of Saudi Aramco's budget, but the other sectors normally relied on government budget allocations. It was therefore sensible to attract foreign investment to those sectors and free up funds for other public expenditure. As an Aramcon participating in the discussions explained, "We wanted to leverage the upstream involvement." Perceptions of this bundling proposal were thus initially positive, but arriving at an agreement with the companies on the terms of the bundled investments became quite difficult. According to Saudi Aramco, the largest of the core ventures, Core Venture 1 (which was headed by ExxonMobil and included Shell, BP and ConocoPhillips) expected an 18 percent rate of return for investing in utilities, which Saudi Aramco felt would normally fetch an 8–9 percent rate of return. This understanding was later contested in interviews with an executive from one of the Core Venture 1 companies. Though Saudi Aramco focused on the rate of return, the more crucial consideration from the foreign oil companies' perspective related to the general commercial viability of the project and to the eventual cost of power and water sold in the Saudi market. The "prospectivity" potential (that is, the potential for discoveries) for the area on offer was deemed to be inadequate by the companies, which worried that exploration and development would not yield enough gas to feed into the bundled petrochemical, power and utilities projects.

As the negotiations dragged on, frustration mounted on all sides. It was a classic problem of mismatched expectations. The investing companies did not want to be cornered into a bad deal. They had been invited to develop upstream gas assets in the kingdom, and they expected "a risked potential for" attractive gas assets to be on offer (which means a potential for discoveries with an understanding that there are no guarantees). They also wanted to be treated as long-term investors rather than as contractors. For their part, Aramco and the Ministry of Petroleum and Mineral Resources were keen to negotiate a deal that would be to the advantage of the kingdom and to impose their views on the negotiation process. In addition, Saudi Aramco wanted to defend its special status and its reserved gas acreage and to dispel any IOC hopes of gaining access to the country's oil reserves. It saw itself as the guardian of Saudi Arabia's hydrocarbon resources. Therefore, the decision of ExxonMobil's CEO to meet only with the crown prince, bypassing Saudi Aramco and the ministry of petroleum, was perceived as an insult by many Aramcons. Mr. Raymond may have been invited to follow this protocol by the kingdom's top leadership, but it is probable that Aramco feared this apparent snub would have repercussions for its role in the kingdom's power relations.

Indeed, influential members of the royal family did want to break the company's iron grip on the hydrocarbon sector. Prince Saud Al-Faysal was a moderate reformer, who, like the crown prince, wanted to present Saudi Aramco with a degree of competition in order to stimulate the gas sector, though this was never meant as a fundamental challenge to Saudi Aramco's primacy. But rival princes and officials from poorer ministries, who sought to benefit more directly from the resources of the hydrocarbon sector, were politically active in the background. In the meantime, the arguments of Saudi Aramco and the Ministry of Petroleum and Mineral Resources were picked up by elements of the royal family opposed to the crown prince's drive for economic reform—notably the surviving members of the "Sudeiri Seven," the sons of the favorite wife of the first king of Saudi Arabia.[9] In this context,

9. Surrounding King Abdullah in the senior councils of the royal court was a clique of seven full brothers, referred to as the "Sudeiri Seven" after their common mother, Hassa bint Ahmad Al-Sudeiri. These brothers are now six, as Fahd was among them.

a manager from Saudi Aramco explained, its objections to the demands made by the IOCs were criticized by groups in government critical of its special status as deliberate obstacles to the plan to attract foreign investment, "and that brought havoc to the situation." The negotiations on the terms of investment became entangled in political rivalries within the royal family and between Saudi institutions. According to some Saudi Aramco executives, ExxonMobil inflamed these tensions by leaking information to government circles about Aramco's "disloyal opposition."

From interviews, it appears that Aramcons were indeed quite forceful in ensuring that Saudi Arabia got a good deal from the foreign oil companies. Interviewees' comments focused on the terms of investment. One executive explained that there was tension in the company about the timing of the investment. As the discussions became further protracted, the initial period of low oil prices, which had motivated the crown prince to seek out foreign investment, gave way to higher prices. Saudi Aramco felt that Saudi Arabia would now get a bad deal. In the end, as a government official explained, "Saudi Aramco and the ministry agreed. The costs were too high." The view from Core Venture 1 was that the deal was not sufficiently competitive in the absence of a substantial gas resource potential. The long-discussed bundling of core and non-core activities was abandoned.

In the two further rounds of the gas initiative, discussions continued regarding the best means to receive and evaluate bids from interested companies, to grant licenses and to regulate the performance of the industry. The foreign minister's preference for negotiations meant that companies would be selected on the basis of qualifications, before making the financial and technical terms of their planned investments known. Aramco lobbied for a bidding process through which all qualifying companies would simultaneously submit their technical and commercial offers for consideration by the relevant government institution. The foreign minister's choice was imposed on the process in the first and second rounds of the gas initiative, "but Saudi Aramco got the [government to accept] bids for the third gas initiative," according to an influential Aramcon.

Then came the issue of who would handle the bidding process and grant the licenses and who would audit the performance of the IOCs. The SPC initially considered establishing a regulatory body, separate from the oil ministry, but it decided that the Directorate General for the Ministry of Petroleum and Mineral Resources in Eastern Province (which the minister chairs) would be the regulator. Saudi Aramco would therefore not oversee the international oil companies. A ministry official commented, "Regulation is (and should be) part of the ministry. The gas initiative will have 'an army' regulating the foreign companies." My impression was that Saudi Aramco had won the battle with the foreign minister over the licensing processes: investors in the gas initiative would be required to show their cards before any promises were made to them. As a result, it could step back and let the Ministry of Petroleum and Mineral Resources take over.

Kuwait

The transfer of licensing and regulatory responsibilities to the oil ministry is a trend in other producing countries too. Kuwait's Ministry of Energy has sought to take on these responsibilities. So far, KPC and some of its subsidiaries have exercised the prerogatives of a licensing regulator. Moreover, at the time of interviewing in 2004 it appeared that it intended to regulate the work of IOCs selected to develop the northern fields in Project Kuwait. An executive at KUFPEC critical of the situation explained that KOC was regulating the industry on aspects that it did not itself fully grasp. "KOC realized it couldn't do the job [of expanding capacity in Kuwait's northern fields], so it said it needed external help. Now it's put in charge of choosing the right technology and the right investor for a project which is beyond its means and of regulating the company that does the work—a work it does not know how to do!"

It was apparent that Issa Al-Own, the undersecretary for oil, was attempting to recover some of those functions. In 2004, he announced that the ministry would execute a new oil strategy in 2005, one that focused on "supervising and completely monitoring the oil sector." It was reported in September 2004 that a ministry team would conduct field inspections and coordinate with executives of the national oil companies in order "to enhance performance and realise defined strategic goals."[10] In Kuwait, as in most producing countries, the ministry does not have sufficient expertise to play the role of regulator effectively, but plans to train its officials for the task. Building this expertise is an important element of reform toward a regulatory model, but it is questionable whether that expertise should lie within the oil ministry. As a company manager aptly put it, "It is inappropriate for the ministry to set policies and then judge the performance. The regulator has to be deeply involved in the industry. It has to understand the details of the operations to judge how the project is going. That is not the expertise and the job of the ministry people. These people are strategists. They have to take a step back and see the big picture for policymaking."

10. Kuwait News Agency (KUNA), 20 September 2004.

It appears that in a strategy document prepared for the oil ministry, Issa Al-Own also claimed for the ministry a number of other policy functions that were held by the company. These included representing Kuwait at OPEC meetings and setting and managing quotas, which are normally key ministry prerogatives. These had left the ministry when Seham Razzuqi departed to take up her post as Managing Director of Financial and Administrative Affairs and International Relations at KPC. It is likely that these functions will be more easily reclaimed by the ministry now that she has left KPC. The ministry is also pressing to take back responsibility for the joint development area with the Saudis, that is, to manage the operator of this development, which is not owned by KPC. Such functions came to fall under the responsibility of KPC over time, as a result of a perceived lack of capacity or direction at the ministry. Clearly, one reason for this was that it has been weakened by frequent changes of minister (there have been nine since 1990). However, some KPC managers felt that the ministry had succeeded in increasing its technical competence by taking on new responsibilities.

The separation of roles is perceived as a very important issue in Kuwait, and it was mentioned in almost every interview in the oil ministry and in KPC and its subsidiaries. There is an apparent muddle of operational, policy-making and regulatory roles between KPC, the Supreme Petroleum Council, the ministry and parliament. It seems that the Kuwaiti oil industry needs a system that makes clear the specific responsibilities of each institution. Certainly, there is support within the institutions concerned for clarifying the lines of responsibility. Kuwait's Supreme Petroleum Council wants to remove overlaps in missions between the company and the ministry. And senior figures in KPC agreed that the policy and operational roles should be divided between it and the ministry, even though this would involve transferring some responsibilities to the latter. However, a top executive felt it was crucial that the company should retain decisionmaking authority over operations. At present, its operations are hampered by excessive interference from state institutions. Fear of corruption motivates much of the stringent Kuwaiti auditing procedures. There is also a common assumption that people down the chain cannot make sensible decisions and therefore must be supervised carefully. As a KPC finance director explained, "The division of finance between the state and the NOC is fine. The division of roles is a problem. . . . We have accumulated funds and therefore we have a high degree of financial independence. We don't have much debt on our books. The problem is the ability to make decisions. Anything that's above KD5,000 [$17,000] has to go to the CTC [Central Tenders Committee] for tender. That's not new. What's new is that anything above KD100,000 [$340,000] must go to the audit bureau and to the CTC."

Iran

In Iran, the blurring of boundaries between Ministry and NOC is legendary. During my visit, I was often confused about the exact position and affiliation of the person I was interviewing. When I was due to meet a manager of NIOC, I would find him in a ministry building, and a ministry official would be working in an NIOC building. Iranians joke good-naturedly about it, but their system is a maze that foreign investors may find impenetrable. As one interviewee pointed out, "Now there are multiple posts, numerous people in charge: the minister, NIOC managers, the advisers. Now, whom do you listen to? Who is in charge? One of these guys expresses doubts regarding the buy-back and then of course Total loses faith and loses interest." In an industry of such importance, there are several centers of power, and oil policy in Iran currently suffers from a multiplicity of "official views." The picture is made more complex because decisionmakers on Iran's energy scene are multiple post–holders, and there are numerous crossover links between the ministry and the boards of the NOCs and their subsidiaries.[11]

Until a few years ago, the Minister of Petroleum was the chairman of NIOC. Today, the minister still holds overall power. The Ministry of Petroleum and NIOC work so closely that it is difficult to distinguish between them, and an adviser explained that some ministry employees are paid by NIOC. The company is clearly under the control of the executive government; and a circle may be more accurate than a triangle as a representation of the distribution of roles in Iran's oil and gas industry—a circle shared by NIOC and the executive, with other institutions, such as parliament and the Guardian Council, acting as centrifugal forces outside it.

However, there have been signs that the clarity of the state–NOC relationship might improve in Iran. The seventh parliament recently reviewed the second introduction of a bill for the reform of state–NIOC relations, which includes the controversial removal of the energy subsidy system and proposes a new structure of relations between the state and NIOC.[12] However, these

11. The Minister of Petroleum is the Head of the Board on the four major companies, and the deputy ministers are the head of the board for the subsidiaries. For instance, the Deputy Minister of Petroleum is the Managing Director of the subsidiary National Iranian Oil Refining and Distribution Company (NIORDC).

12. The sixth parliament passed the bill but it was then blocked by the Guardian Council. As a result, the bill was sent back to parliament for review of specific elements of the bill. It was

reform plans are caught up in a power struggle between parliament, on the one hand, and NIOC and the Ministry of Petroleum, on the other. Members of parliament want NIOC to include in the bill new articles that set out how the company is to interact with the energy ministry and other ministries (*MEES*, 18 July 2005: 29). The company's reluctance to do this is presumably tied to its concern that parliament will then amend the articles in order to increase state control over the company and parliamentary control over oil and gas revenues.

Parallel to these discussions, NIOC commissioned a study to advise on the fundamental reorganization of its structure and its relations with the state. It hoped that the recommendations of this study (including those on finance, strategy and operations as well as the relationship between it and the government) would be implemented in phases beginning in 2005. This internal reform and the reform bill being discussed seek to make the company more commercial and to increase its productivity. An influential adviser involved in the study explained that NIOC had to find ways of increasing its available capital because the government is running deficits. It appears from interviews that the plan to restructure NIOC was supported politically by the Minister of Petroleum at the time, Bijan Namdar Zanganeh, and others in the government. But some closely involved were already concerned in 2004 that the "political will isn't there" to see through the restructuring.

High crude oil and oil product prices help the push toward reform because gasoline imports are very costly, but reformists worry about the new parliament and whether it will block the process. The recent election of Mahmoud Ahmadinejad to the presidency casts further shadows on the oil industry and its hopes for the proposed reform. The new president has given signs that he plans to subject the industry to greater central control. He is expected to fight against corruption (including political nominations) in the industry, and he may question foreign investment in the sector.

Abu Dhabi

The three-layered cake in the Emirates appears to work smoothly. There are no apparent difficulties owing to overlapping roles and responsibilities between ADNOC, the government of Abu Dhabi and the institutions at the federal level. The Ministry of Energy does not play a real balancing role vis-à-vis ADNOC because it functions at the federal level, whereas the management of

---

reviewed by the seventh parliament. A number of objections were raised regarding Articles 3 and 4. Discussions are under way.

the oil industry is the prerogative of each emirate. Recent changes at the ministry level were not designed to modify the power structure. The newly integrated Ministry of Energy (responsible for electricity and water), like the old Ministry of Petroleum and Mineral Resources, has limited power to set policy and engage in overall planning. The main function of the oil minister is to represent the Emirates at OPEC. Policymaking is the responsibility of the Supreme Petroleum Council, on which the minister himself does not sit. However, the new minister, Muhammad Al-Hamili, will have more influence within ADNOC itself because he is the chairman of the Abu Dhabi National Tanker Company and the director of marketing at ADNOC-Distribution. Nonetheless, the ministerial post remains a "political position," as one ADNOC manager put it, and these recent changes are unlikely to have an impact on the role of the energy ministry. As an upstream manager said, "The government is the driver of policy," meaning the ruler of Abu Dhabi. The center of power will remain with the SPC, which manages ADNOC and is chaired by Sheikh Khalifa (who succeeded his father on the throne), and with the General Manager of ADNOC, who "has both keys." But as an ADNOC manager commented, "It's all one body. In all government, there are no contradictions in decisions." The Abu Dhabi system was extraordinarily centralized in the person of Sheikh Zayed, and this system gives all the signs of maintaining itself under the new leadership of Sheikh Khalifa. As a result, there is a great deal of unity and stability in the decisionmaking process. The following comment from an upstream executive highlights this: "I can talk to you [about the policy] in spite of the change in leadership because I know it is stable. I know that things won't change. It's the same philosophy, the same policy."

Conclusion

Improving the governance of the oil and gas sector is a high priority for governments and national oil companies in the Middle East and North Africa. This was demonstrated in the reasons given for allowing me to conduct interviews: NOC and oil ministry representatives expressed the hope that the study would give them an external insight into the quality of the relationship between state institutions and the NOC. Discussions revealed that in Kuwait, Algeria and Iran, the national oil companies have been trying to increase their operational autonomy. The ministries of petroleum in Kuwait, Iran and Saudi Arabia have asserted their responsibility for policymaking and regulation, and the Algerian ministry has given that prerogative to a regulatory agency. Parliaments are also playing a greater role. These redefinitions of the balance of power and the functional division between the national oil companies and state institutions may bring some confusion during the transition, but they are a necessary adaptation of governance processes to the changing political landscape. State institutions are stronger now in all these countries and are better able to play an effective role in policymaking and regulation. Meanwhile, the companies are more commercially driven, and they assert their needs in terms of setting up clear decisionmaking processes that enhanced performance. Generally, it appears that in cases where the political leadership has sent clear signals on the demarcation of roles and responsibilities in the oil sector, the governance of it has improved.

five
New Generations, Changing Expectations

As we saw in Chapter 1, the Third World producers' realization of the power of their resources and the parallel weakening of their rivals the Seven Sisters led to the emergence of a new world oil market. There evolved a progressive challenge to the privileges of the foreign oil companies through the creation of OPEC and the wave of nationalization of oil companies in the early 1970s. In these ways, the Middle Eastern oil producers were taking back and developing their resources and establishing their importance on the regional and international scenes.

Regionally, the windfall of oil revenues that followed the oil crisis of 1973 and the fall of the Shah of Iran in 1979 propelled the Gulf monarchies to a position of influence. Saudi Arabia and Kuwait became powerful benefactors in the Middle East. Led by Riyadh, the Gulf producers created the Arab Economic Development Fund and ushered in an era of Arab economic solidarity. This had the effect of weakening the political grip of Egypt and Syria on the region: solidarity was now commanded less by the socialist-nationalist slogans of the Nasser era and increasingly encouraged by state aid and employment opportunities in the Gulf. Support from petrodollars spread economic dependence on oil revenues throughout the region. For instance, Jordan, which produces no oil, was dependent on foreign aid and loans from the Gulf for 43 percent of its public earnings between 1984 and 1990 (Brand, 1994: 48).

The Political Contract and Popular Expectations

The whole of the Middle East engaged in projects for industrial and economic development funded by oil revenues and supported by subsidies. The "rent" (a high level of government earnings that are not from taxes) from oil exports conditioned the industrial, investment and development profile of these countries; it suffused the political system and the relations between state and society. The political significance of oil to the Middle East and North Africa was not limited to the empowerment of national leaders in the face of former colonial powers. Oil was now the building block of these countries as modern states.

Following the wave of nationalizations throughout the region, states became the primary agents of the economy and the generators and distributors of the oil rent. As "rentier states," oil exporters were characterized by the capacity of the state to distribute large amounts of financial resources to society—resources that did not come from taxes and therefore gave the state a degree of autonomy from society (Brynen, 1992: 74). The state was no longer dependent on direct taxation. The level of state reliance on public resources from this "exterior" rent is an indicator of the rentier nature of the state; it is usually set at 40 percent (Luciani, 1990: 76). The five countries under study easily meet these criteria, with dependence on oil export revenues ranging from 60 percent in Iran to 81 percent in the UAE (see contribution by John Mitchell at the end of this volume).

Rentier politics occur in a variety of political regimes (monarchies, military, socialist and revolutionary regimes and parliamentary republics). Though each country has a specific political context, a common characteristic of all their politics is that the state's legitimacy has been built around its capacity to distribute rent to different segments of society. In Saudi Arabia, for instance, the existing taxation regime was replaced by a distributive bureaucracy that used the allocation of oil revenues as a tool for controlling the economy and society. As the political scientist Kiren Aziz Chaudhry has explained, "This freed it from a debate about its collective goals and limited public pressure so long as 'a baseline threshold of legitimate expenditure' was passed" (Chaudhry, 1997: 33). And this, according to her, led to political apathy.

Rent is distributed not only to a regime's allies but also to its potential challengers and the poorer segments of society, because dependence on the transfer of resources builds allegiance to the status quo. States commit themselves to the regular allocation of subsidies (including the subsidized pricing of energy for domestic consumers) and transfers to specific social groups and to administer free public services, to provide jobs in the public sector, to minimize taxes, and so on. In exchange, the various sociopolitical groups give their allegiance to the regime without making demands for political participation. (In the case of Saudi Arabia in the boom years of the 1970s, the state created a new economic elite, a Najdi private sector, and minimized its dependence on alternative, largely Hijazi, elites via the distribution of oil revenues. In the process, the new private sector became dependent on the government, which dominated the economy.) But conversely, close links between members of the government and the business community can make it difficult for the government to introduce reform when state revenues are constrained because reforms might undermine the interests of the state's allies.

The political pact of the rentier state eroded in many producing countries when oil prices fell. The oil price collapse of 1986 exposed weaknesses in many economies, with the immediate effect that governments incurred large internal or external debt in order to try to maintain the rentier lifestyle with borrowed resources. During the 1990s, governments in one country after another began to address the weakness of the non-petroleum sectors of their economies—in some cases under duress of International Monetary Fund (IMF) programs to which their foreign borrowing had exposed them. The current spell of high oil prices has lessened the sense of urgency with which exporting states engage in economic reform. However, as producers themselves explained, these price trends are unlikely to continue in the long term. John Mitchell's analysis in his background contribution at the end of this volume demonstrates that scenarios of more modest prices will lead to a declining ability of the state to maintain the level of support it has given its citizens in the past. Demographic trends are an important part of these equations, as governments in the region are well aware. This chapter will describe the demographic trends that have inflated demands on the state's rent and changed the expectations of society. Where relevant, the analysis will illustrate how these trends have affected the national oil company.

New Generations

Two demographic trends are changing the face of the Middle East, in particular that of the Gulf states. These trends are bound to have a profound impact on not only the economies and societies of the region but also the operating environment of the national oil companies. The first trend concerns the emergence of a new generation of the ruling class. The generation that came to power on the heels of the colonial powers is aging and giving way to the next generation, of 35- to 50-year-olds. Political successions have become commonplace: in Qatar in 1995, Sheikh Hamad bin Khalifa Al-Thani overthrew his father and seized power at the age of 45; in 1999, King Abdullah of Jordan took his late father's throne when he was 37; in Bahrain in 1999, Hamad bin Isa Al-Khalifa succeeded his father at the age of 48; and in 2004, after the death of Sheikh Zayed, the man who was the architect of Abu Dhabi, the throne passed to his son Sheikh Khalifa, at 57 slightly older than the new crowd. However, in Saudi Arabia the succession system is not making room for the younger generation: power is transmitted laterally from older brother to younger brother rather than down from father to eldest son.[1] King Abdullah and Crown Prince Sultan are both in their eighties, and other princes in the line of succession are equally aged.

The new generation of rulers is more pragmatic and less ideological than their fathers. Even so, they have retained a number of lessons from the nationalist period. This is reflected in the attitudes of the executives managing the region's oil industry. They have inherited many aspects of the traditional value system but were raised in modernity. In Saudi Arabia, for example, the 35- to 50-year-olds are part of a new, educated and urban middle class that maintains some traditional attitudes and follows a pattern of patriarchal authority transmitted by the older generation (Yamani in Hollis, 1998: 145).

Born between 1955 and 1970, this generation's youth was set in relatively modest conditions. They did not grow up with consumables such as automobiles and televisions. They discovered wealth with the oil boom. In the Arab countries of the Gulf, opportunities have been plentiful for this generation, many of whom were born in rural areas or in the desert and moved to the cities to acquire an education and enter the workforce. With the rapid development and modernization of their countries, they were able to rise quickly through the ranks. Similar opportunities were available to those of their generation in Iran and Algeria, though political upheavals and difficult times in those countries shaped their lives quite differently.

The 35- to 50-year-olds look on the next generations with a mixed sense of pride and concern. These young people make up the second shift in the region's demographics. The "baby boom" that followed the 1970s oil boom has produced a large generation of 20- to 34-year-olds. Population growth

1. This precedent was set by King Abdul Aziz, the founder of the monarchy, in 1953. It was meant to forge a union between two of his sons, Saud and Faysal.

rates rose steadily, and the following generation, those under 20, now dominates the demographics of the Gulf. In Iran, for instance, 32.6 percent of the population was under 15 in 2002 (UNDP, 2004: 154). Throughout the Middle East's oil-producing states, the under-20s make up over 43 percent of the population on average (see figure 5-1).

These new generations clearly have different aspirations and values. They were raised in a world different from that experienced by the previous generation. They are more educated, urban, consumerist and, thanks to the telecommunications revolution (the Internet, satellite television), more open to the world than their parents. But their greater awareness of global trends has not necessarily brought adherence to the Western value system, even though they enjoy access to more diversified sources of information and express themselves more openly than previous generations. The Saudi sociologist Mai Yamani explains that though the ruling royal families of the Gulf still refer to varying degrees to their Islamic and tribal values as a way to maintain their political legitimacy, the generation that has come of age in the past decade has been removed from that pastoral and nomadic past. Their fathers carry the memory of desert tribesmen, but they know only a world of urban and modern lifestyles funded by the income from oil exports (Yamani in Hollis, 1998: 136–48). The rising generation, coming of age in the next 10 years, knows another world still. They are experiencing the particular strains that result from increasing awareness of the vulnerability of their societies to declining oil revenues.

The demographics of the Middle East and North Africa indicate rapid population growth, as figure 5-2 shows. By 2015, it is projected that Iran's population will have doubled since 1980, to nearly 80 million (UNDP, 2005: 233, and *UN Demographic Yearbook 1986*: 178). Although Iran's population policies since 1988 have reduced the natural rate of population increase to around 1.4 percent per year (Abbasi and others, 2001: 1, 8), the effect of higher previous rates is feeding through into the cohort of those of young working age, and there is also a much greater expectancy of a longer life. On the other hand, more young people are expected to remain in education after the age of 14, so that the increase in the labor force in the next decade will be around 4 percent per year.[2] Many Iranians spoke to me about the emerging power of young people. It is said that they are better educated and want a say

2. This is according to the *World Bank Country Brief: Iran*, September 2004, available at *http://siteresources.worldbank.org/INTIRAN/Overview/20195737/IRANBRIEF%202004AM.pdf*. For a review of the development of the agenda of economic and social reforms through the five-year planning process, see Farjadi and Pirzadeh (2003).

Figure 5-1. Percentage of Population under 20, Selected Countries, 2005

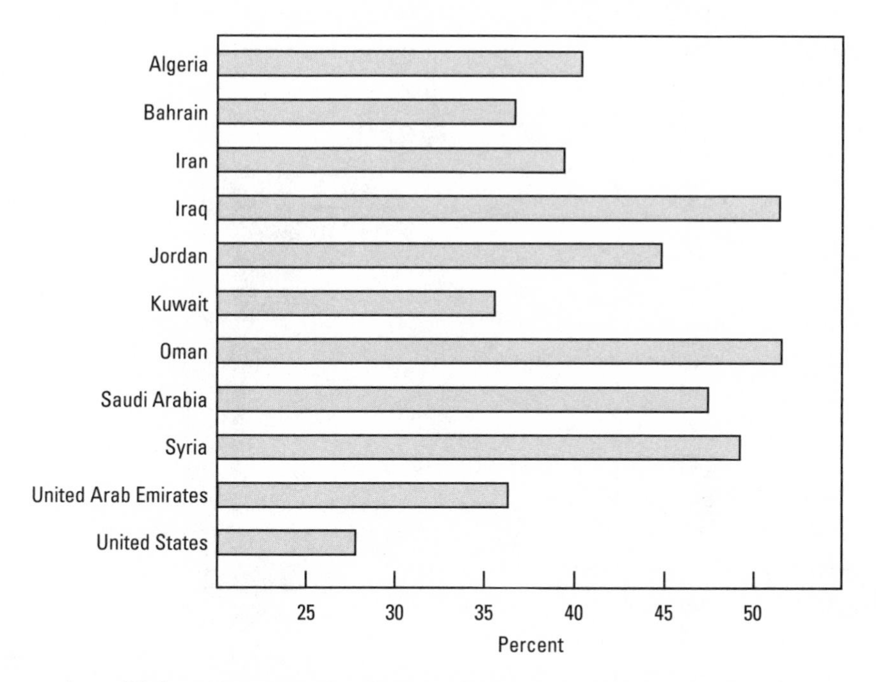

Source: U.S. Census Bureau International Database estimates, http://www.census.gov/ipc/www/idbpyr.html (accessed November 2005).

in how the country is managed. My host in Iran felt that they would be a force to contend with. It is unclear how (and in what numbers) young people voted in the presidential election that brought the conservative mayor of Tehran, Mahmoud Ahmadinejad, to power in June 2005. However, it would seem that they are not yet engaged in Iranian politics.

In contrast to Saudis, those Iranians entering the job market every year will find more opportunities because the economy is more diversified. Employment increased from 14.8 million in 1998–99 to 18.7 million in 2003–04, absorbing the reported increase in the active population so that unemployment remained constant at about 2.5 million, bringing its reported rate down from 16 percent in 1999–2000 to 11 percent in 2003–04. However, these opportunities are neither attractive enough nor sufficient to prevent a devastating exodus of professional people. The IMF has estimated the number of educated Iranians leaving the country each year for better opportunities at 150,000, making it the world's worst country for brain drain. The national

Figure 5-2. Population and Projected Population: Algeria, Iran, Kuwait, Saudi Arabia, the United Arab Emirates and Iraq, 1980–2020

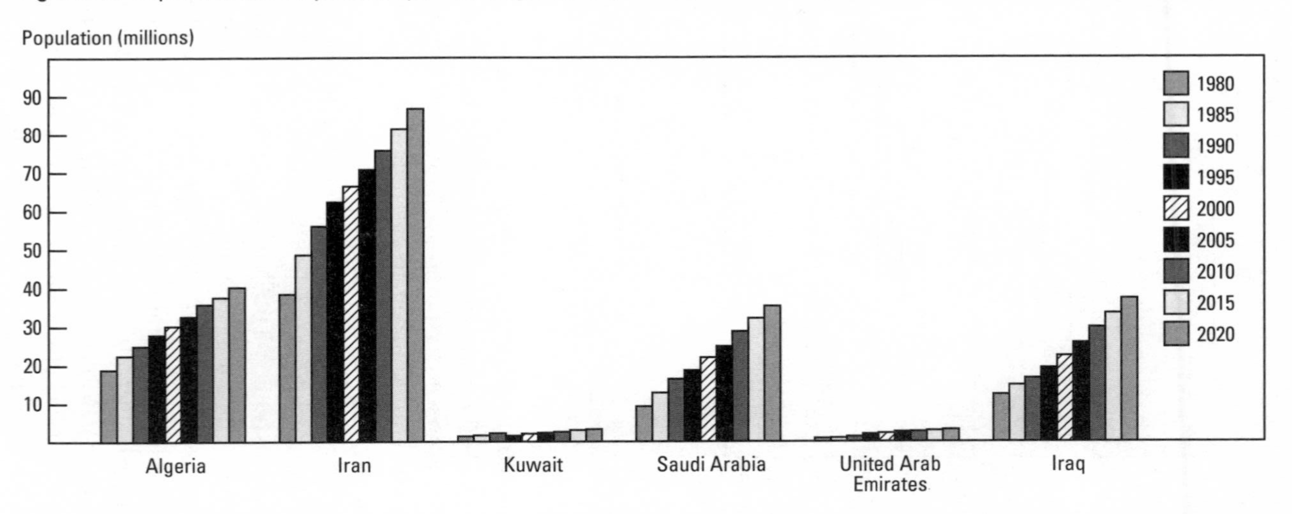

Source: Population Division of the Department of Economic and Social Affairs of the United Nations Secretariat, "World Population Prospects: The 2002 Revision and World Urbanization Prospects: The 2001 Revision," http://esa.un.org/unpp (November 9, 2004).

petroleum sector itself faces competition for skilled employees within Iran because the salaries it offers cannot compete with those offered by foreign private investors in the country.

Education

The education of women in large numbers is a compensating demographic trend that is likely to transform the social and economic landscape of Iran. More women go to school and stay in school (in 2001, women entrants exceeded men at Iranian universities), so that female participation in the labor force is expected to rise from the current low level of around 15 percent. Women make up over 60 percent of university graduates,[3] and they dominate enrollment in engineering schools. As a number of managers of the National Iranian Oil Company have pointed out, they should come to take an important role in the future management of Iran's oil industry.

It is a largely ignored fact that the Islamic revolution of 1979 gave Iranians greater access to education by making it more accessible to less privileged classes and by increasing the enrollment of girls. The increase in their enrollment was in part due to the Islamization of the country's educational institutions, which encouraged traditional and conservative families to send their daughters to school. The schools were no longer mixed, and this permitted more girls to attend. Meanwhile, these changes did not deter more secular, middle class families from sending their daughters to school.

In the Gulf, higher education rates are not producing the desired economic growth. Diwan and Girgis (2002) explain that in Saudi Arabia, the impressive increase in the average extent of the education of both sexes by 1.5 years to 6.6 years in the past decade has not led to the expected corresponding growth of the economy. The young are more educated than before but find fewer jobs. This is due to the lack of productivity of the non-oil sectors in Saudi Arabia and its neighbors, as John Mitchell's analysis demonstrates in his contribution at the end of the book. But clearly there is also a mismatch between the skills of nationals entering the workforce in those states and the needs of the private sector.

The United Nations Development Program's Arab Development Report explains that across the region, higher education systems respond weakly to the labor market needs related to science and technology. To take Kuwait as a case in point, 44 percent of university graduates receive degrees in

3. *http://www.parstimes.com/women/women_universities.htm.*

humanities, 18 percent in public administration and 12 percent in Islamic law. Only 26 percent take a degree in health and science. The report finds that the percentage of students enrolled in scientific disciplines in higher education is small compared to advanced countries such as South Korea (UNDP, 2003: 71–72). This is partly a result of the control of the education system by conservative religious authorities in countries such as Saudi Arabia, Kuwait and the UAE. The governments have succeeded in raising the rate of school enrollment to 90 percent but they have let the school syllabus be dominated by religious studies—even in technical degrees (Dazi-Heni, 2001: 10). In Saudi Arabia, this policy dates back to the 1970s. The government encouraged a conservative tendency by giving the religious establishment (*'ulama*) supervisory control over educational policy, even in the secular universities, resulting in a strong Islamic influence on the curricula. The teaching of foreign cultures and beliefs was restricted, and foreigners were excluded from state schools.

The education system of Saudi Arabia has expanded greatly since then, but it does not appear to integrate the diverse social groups of the country equally. The rural and urban lower classes tend to drop out. Also, critics point to the failure of the system in terms of the quality of education. Saudi youth are competing in a sophisticated market with underdeveloped skills. This could cause frustration and resentment for young graduates, which can only be compounded by the decline of state aid. Their inherited deference to patriarchal authority means that the young blame the employers and turn to the government for solutions. The youth apparently resent the realities of a global marketplace. They express their dissatisfaction with the need to be mobile (to leave the region to which their tribe belongs in order to get a job) and with language requirements (usually English) because they are trained in an environment stressing the value of Arab and Islamic communal values (Yamani, 1998: 145).

In interviews, a number of Saudis expressed the view that the education system should encourage relevant skills rather than focus on religious teaching. A number of young women in Saudi Aramco felt that religious teaching was overly emphasized, as this comment indicates: "Both girls and boys get a religious education. But in the newer generations, girls are more conservative than the boys. We must find a balance between the conservative and the modern. We want them to learn religious principles, but we don't want them to be trained as theologians!" Those who had experienced national and international schools expressed the need for a national education system that encouraged creativity and innovation in students.

Education systems affect the attitudes of personnel in the national oil companies. NOCs need to recruit nationals who have a quality education in a professionally relevant field and have been exposed to international thinking and standards. In interviews, most people felt that national universities trained students to high educational standards, especially in the "classic" fields. However, those trained abroad felt that international experience was very important. One young Saudi told his story. He graduated at the top of his class from King Fahd University of Petroleum and Minerals and was recruited to Saudi Aramco. He felt very confident about his qualifications until he worked with other recent graduates who had been trained in American and European universities. He perceived his own qualifications as markedly inferior and resolved to pursue further studies abroad. When his request for leave was refused (on the grounds that his position could not be held for a year because so many young graduates were eager for it), he threatened to quit his job in order to demonstrate how important this was to him. He won his battle and successfully completed an MBA program in Britain.

Employment Opportunities

Concern about the employment situation does not affect all states equally. Abu Dhabi, for instance, is quite different from the oil producers with large populations such as Saudi Arabia and Iran. This small emirate is training its youth at an accelerated pace so as to provide the oil industry with qualified replacements for retiring expatriates. A worry there is the small national labor force. In Saudi Arabia, Algeria and Iran, the employment challenge is far greater, as a new tide of educated applicants floods the job market every year. Manifestly, the oil industry cannot offer enough jobs.

In other Arab Gulf countries besides Abu Dhabi, governments are striving to "nationalize" their workforce, to replace retiring expatriates with nationals. "Nationalization" policies have been successful in state enterprises, specifically the national oil companies: Saudi Aramco has reached 86 percent in the "Saudization" of its personnel. Elsewhere, however, the results are unimpressive, with an average decrease in foreign workers in Saudi Arabia over the past five years of only one percent per year. Private-sector employers are reluctant to employ nationals who are not part of their family. The challenge to "nationalization" is essentially the availability and competitiveness of the foreign labor force. Unofficial figures set the proportion of foreign migrants in the kingdom's workforce at 65 percent. This share is naturally higher in Kuwait and Abu Dhabi, where the small populations give few alternatives to dependence on relatively cheap labor. Though these states depend on the availability of this labor for the functioning of their non-oil sectors, they need to upgrade non-oil activities into higher value-added businesses that can afford to match the salaries offered to nationals in government service.

Meanwhile, governments have privatized a number of state companies, the traditional employers of nationals, and they have had to streamline their operations and lay off employees and civil servants. The overflow of job seekers cannot be absorbed by the state. And private-sector opportunities for nationals are limited. This situation has provoked dissent in countries such as Oman, Bahrain, Kuwait, Saudi Arabia and Iran since the 1990s.

Changing Expectations

The young people joining the labor force now are ambitious and expect rapid promotion and plentiful opportunities. But in fact, their professional prospects may not be as good as those of their parents, who rose quickly through the ranks to fill the shoes of departing foreigners. This difference is readily recognized by the older generation. There is growing concern about the unemployment rate, which is particularly high in the 20–34 age group.

The political discourse of the ruling generation appears to be preparing the youth for the deep transformations of society. This discourse is not from the ruling class's political heritage of nationalism, decolonization and the Cold War, which predominated from the 1960s to the mid-1980s. It is set, in theory at least, in globalization, liberalization and structural adjustments. The new generation of leaders is subject to the pressures of the World Bank, the International Monetary Fund and the World Trade Organization. They are told to reduce state subsidies, to privatize branches of the public sector, to streamline the civil service, to modernize the administration and the economy and to impose taxes. In other words, they are asked to dismantle the privileges granted by the previous generation of rulers to the various groups of society.

In speeches to their young nationals, the new rulers in the Gulf often caution that the golden age belongs to the past. Since the sudden fall of oil prices in 1986, the generation that is today between 20 and 34 years old has felt itself caught between the expectations of wealth that came with the boom of the 1970s and a more somber economic reality. Not only did oil revenues plunge and then stagnate with the downturn in prices in the late 1980s but demographic growth also put the constrained oil rent under greater pressure. While production has basically been flat in the past decade in countries such as Saudi Arabia, Kuwait and Iran, the population has increased significantly, rising by over 25 percent in Saudi Arabia and 18 percent in Iran since 1990. Accordingly, per capita GDP during this time has been stagnant in Saudi Arabia and Kuwait and has declined by about 10 percent in Iran. Continued high oil prices in the coming years are uncertain. In a context of growing unemployment, young Saudis, Kuwaitis and Qataris, though more educated than previous generations, cannot find employment in line with their ambitions. Demand clearly outstrips available management positions. They would naturally turn to the state for respectable positions, but even the state cannot meet demand without mortgaging itself with large deficits. Unofficial unemployment figures are usually estimated to be 20 percent.

In the Arab countries of the Gulf, the older generation felt that they had had it easy, whereas the new generation was facing harder economic times. One Aramco manager felt that it would be necessary for Saudis to shift their mentality from "consumers of wealth" to "creators of wealth." "We must think differently now. That is a real challenge for us and for states like us. When I look at my office, my house, I think, have we created this? No, we have imported it. Importing is not bad in itself, but we have to create things as well. This will bring jobs. It is certainly difficult for this generation. The wealth is declining with the increasing population."

In conversations in Riyadh and Dhahran, a number of Saudis wanted to dispel an enduring perception outside Saudi Arabia of their young people as spoiled and unwilling to take on low-grade positions. I saw an example of the change in mind-set on a flight of Royal Saudi Airlines: all the male crew was Saudi.[4] The company's in-flight magazine boasted that 97 percent of the on-board male staff of the airline are Saudi. In a society that has undergone a halving of the per capita income in just two decades, from a peak of $15,999 in 1980 to $8,373 in 2002,[5] ambitions have fallen closer in line with economic realities. The number of poor Saudis has increased, and they are visibly crowding the streets of the major cities. The kingdom only acknowledged this as a problem in 2003, after the then Crown Prince Abdullah made a surprise visit to a poor area in Riyadh in November 2002. He launched a major survey of households in 2004 as part of a government campaign to eradicate poverty.

4. In 2001, the airline announced that Saudi female on-board staff would be able to operate within Saudi airspace only (*Al-Hayat*, reported in *Islamic Voice*, February 2001). This is presumably in order to comply with Islamic law in Saudi Arabia, which requires women to travel abroad with their husband or a brother.

5. These figures are given for gross national income (GNI) divided by population at current prices by the UN Statistics Division, *http://unstats.un.org/unsd/snaama/dnllist.asp.*

Those who join Saudi Aramco are the lucky few. The company recruits the best young graduates into its training programs. They are very motivated and ambitious. Though they cannot expect the same rapid ascent through corporate ranks as the present executives of the company, they will nonetheless be rewarded with high salaries and attractive perks. In this respect, NOCs such as Saudi Aramco, NIOC and Sonatrach, which can tap into a large national pool of qualified engineers, have a competitive advantage over the international oil companies, which are faced with a talent gap that could cripple the oil industry. In the West, the oil industry has a negative image and is a hard sell to students, who do not see it as an industry of the future. The average age of oil field engineers is 50 in the United States. But in the Middle East, oil does have a future. The national industry is viewed with pride and often is the best employer. In fact, the IOCs are turning to the producing countries to recruit qualified engineers, offering salaries with which Sonatrach and especially NIOC cannot compete.

By contrast to Saudi Aramco, where young recruits have had to prove themselves in a very selective recruiting process, Kuwait's and Abu Dhabi's national oil companies will face greater difficulties in motivating their recruits. This is because their young people have not yet had to contemplate a significant downgrading of their economic prospects. This is reflected in attitudes at work in KPC and its subsidiary companies, where managers complain about the difficulty of motivating their young staff—even about getting them to show up at work. The work ethic has eroded and management processes have not adapted to address the situation.

The Kuwaiti economy is facing greater challenges than before, most notably in the pressure to provide jobs for the tide of young people entering the job market every year. However, the oil rent per capita remains at a sufficient level to maintain the standard of living, even if it does not increase. Poverty is not yet looming, and expectations of increased standards of living and job promotion remain untouched. For instance, concerned older Kuwaitis told me that there is an alarming level of personal indebtedness among the young. They turn to credit in growing numbers in order to maintain a lavish standard of living they cannot pay for as they try to keep up appearances and to match their neighbor's car. This is also a problem in the UAE, where 90 percent of youth are in debt.[6] This indicates a weakness of the local banking sector, which relies excessively on consumer loans,[7] but it also

6. *Khaleej Times*, 28 July 2002.

7. According to the Central Bank of Kuwait, the annual amount of consumer loans given to residents by local banks rose by 37 percent from 1999 to 2003. The *Lebanon Daily Star*

points to overconfidence among the young about their future economic prospects.

Abu Dhabi faces an economic future similar to that of Kuwait. One executive from a foreign oil company working in the Emirates commented, "There is a worry here. A large part of the expats are from the Indian subcontinent and will soon be retiring. They are running the operations. So [ADNOC's management is] worried about how to replace them. The young people don't have training or experience for that work." So far, high salaries have allowed the company to attract Abu Dhabi nationals to manual positions, but there is uncertainty about the future availability of workers as more of these jobs become available. I asked oil professionals in Abu Dhabi for their views on the change in the new generation's work ethic. A typical response was that "There is a shift . . . Traditions and values are fading away. It's a material world. It's hard to see a picture of the future. The young are very pragmatic. They grew up with everything. Some are lazy. Some don't take education seriously. There is a mix. They are very influenced by Western society but not by the good things [about it]. There is a mixed feeling. . . . "

Interviews with Kuwaiti executives indicated that the new generation has more education but is driven too much by ambition and not enough by work. "The new generation: what they have is English. That's their only asset. They take a lot of time to prepare a nice Powerpoint presentation. Have it all look good. They talk nice, but they don't understand the business, the industry. They want opportunities for promotion. They want to be CEOs. They need training."

"They are more relaxed. They have seen more wealth. And more abundance. They are more educated, have more private schooling. Are they more open-minded? I don't know. They are less committed to work. They get a family allowance, which is [only] a little less than a starting salary!! But they come in with new skills, like IT. They just need to be motivated. They don't see value in what it's all about, in what the company is all about . . . where is it going. Before, in the older generation, we had promotion potential and exposure to the international sphere. [The lack of these things] is frustrating for the new generation. There weren't so many of us at the time and there were more openings. [You could quickly rise and learn new things.] Now, they complain to their boss if they don't get a bonus. They expect it. That's new. Before, we were embarrassed to talk about money. That's the Bedouin culture. We were happy to get anything. To have a salary."

---

reported on 24 December 2003 that between 2000 and 2002, consumer lending surged by 28 percent in Kuwait, by 25 percent in Saudi Arabia and by 20 percent in the UAE.

The perceptions of the younger generation in Algeria and Iran are strikingly different, as are the young people themselves. Throughout the region young employees in the NOCs show ambition and drive, but young Algerians are exceptionally focused. Career and family plans had been put on hold during the long civil war and are now being pursued with vigor upon the return of stability. Algerians do not linger on the hardships they endured; they show mature, cautious optimism for the future. A young woman explained: "Young people in Algeria are very optimistic. There is a lot of potential. They want to work. That's really what people want: to work and to make things happen here. We are able to do a lot with very little."

Some young graduates also feel that they have an edge over the older generation. As one Algerian put it: "I'm worried the older people have trouble adapting to that international focus." He counted on the large proportion of young people in Sonatrach to save the firm from an excessive national focus.

Conclusion

Demographics are changing the producing countries of the Middle East as young people outnumber the old. The growing number of educated job applicants who remain jobless will strain the social fabric and bring increased resentment among the young. Governments, which are bound by the political pact to provide materially for the population, will find it increasingly difficult to do so by depending on the oil sector—that is, the NOCs—to the same degree as in the past. The oil industry in Algeria, Saudi Arabia and Iran will not be able to sustain the needs of the next generation without some support from the private sector. Within the national oil companies of Kuwait and the UAE, the greatest challenge relates to training and recruiting young people to run their operations and to instilling a strong work ethic. The generation of 5- to 19-year-olds being schooled now may grow up with more modest expectations. Employee motivation is not likely to be a concern in the NOCs of Algeria, Iran and Saudi Arabia, where jobs will be harder to come by.

Enormous challenges have to be met in order to sustain economic growth and provide more employment in a future in which sustained high oil prices are uncertain and further growth in petroleum export volumes in some countries will soon become increasingly difficult to achieve because of natural resource limitations. (In the final chapter, we shall examine the current efforts by governments to reform the economies of the five oil-producing states under study.) The future of these countries, and especially of their youth, cannot be secured by oil (or gas) alone and thus the NOCs cannot bear the whole burden of providing for it. The major challenge will be to develop a non-hydrocarbon economy that can prosper in spite of uncertainties and limitations on the petroleum horizon. The NOCs' main contribution will be to continue to add to their country's petroleum reserves and to develop production in the most efficient manner possible. How they have sought other ways to contribute to the national economy will be discussed in the next chapter.

six

Changing Mission of the National Oil Companies

With high oil prices in 2003–05, the governments of Middle Eastern exporting countries are not under immediate pressure to introduce economic reform, stimulate the private sector and reduce public services. However, in view of the demographic trends in the producing countries discussed in Chapter 5, governments will need to foster economic conditions for greater employment opportunities. Short of effective policies to meet these challenges, they may come to rely increasingly on their NOCs to raise their hydrocarbon revenues in line with the growing needs of their populations and their non-hydrocarbon economies. The needs of a society and economy dependent on hydrocarbon revenues range from public healthcare and jobs to foreign exchange to pay for imports. The domestic economic and political challenge of a revenue shortfall could alter the relationship between NOC and state significantly.

Government Pressure on the NOCs

In my discussions with each of the five producers, I tested the proposition that governments faced with increased macroeconomic pressures (and therefore concerned about their political survival in the short term) would intervene increasingly in the management of the hydrocarbon sector in order to maximize the revenues available to the state. I also discussed the risk that excessive interference by government in an NOC's operations and management decisions might in fact reduce its productivity. This interference would be more likely to occur in Algeria, Iran and Saudi Arabia in the light of these countries' economic challenges. It would be less likely in the UAE and Kuwait, which face less pressure in the medium term.

Unsustainable Dependence

In spite of high earnings in 2004 (when most interviews were conducted), it was striking how deeply the above concerns ran in the national oil companies. Interviews revealed that corporate planners and advisers were aware of public budgetary constraints and the potential political fallout from a deterioration of the state's capacity to provide services to the population. An Iranian Ministry of Petroleum adviser, for example, commented that "With the decrease in the standard of living, people will wonder where the money has gone." There is some mild frustration in the NOCs about the state's capability to optimize oil revenues for the welfare of society. A Saudi executive expressed this quite well: "All our savings and all the hard work go into a black box. We never see what happens to those revenues. But it's always like that when the government owns a high-profit resource like oil."

He was alluding to the government wasting resources because of corruption and to the leadership channeling funds through its patronage network. This was rarely discussed in interviews. In Kuwait, the problem of corruption was brought into full view when the managing director of the national tanker subsidiary KOTC won a civil suit in the British Commercial Court against former managers for embezzlement of funds. The ruling detailed how they had, together with a former oil minister, stolen millions from KOTC. Street protests followed the public announcement of the ruling in 1998 (Tétreault, 2003: 91).

Even with a careful management of revenues, NOC managers recognized clearly that the oil rent would not be able to sustain the economy in the future. This problem was discussed mostly in Iran, Algeria and Saudi Arabia. In Algeria, an executive commented, "Sonatrach is the engine of the economy—but it can't support the economy alone." Owing to the sensitivity of discussing political instability in the region, I was particularly surprised at how easily Saudis at the oil ministry and Saudi Aramco spoke about these challenges. The national operating environment is naturally a concern of NOCs, and at Saudi Aramco, it was practically the only challenge that preoccupied managers.

These challenges are, of course, also a concern of governments. To varying degrees, their economic policy is focused on diversifying the economy. This is an important component of government policy notably in Iran, Algeria and Saudi Arabia, where attempts are being made to wean the economy from dependence on the oil rent. The following comments were made in Iran and Algeria respectively:

> Iran: "No one—not the conservatives, the progressives, the executive, the legislature, the clergy, the intellectuals, the young, the old—would want oil and gas to be more important than other sectors in Iran."
>
> Algeria: "Our concern is certainly with the oil curse. We do not want to depend on oil for our economy's future."

Impact on the NOCs

In interviews, I suggested that growing government budgetary needs could have two types of impacts on the NOCs: first, it could curtail their operational autonomy and, second, it could constrain available capital for the hydrocarbon sector. In response, many strategic thinkers in the NOCs felt sure that an economic crisis was looming, but they fell short of predicting a direct negative impact on the NOC.

In Saudi Aramco, senior managers felt confident that the government would not change a relationship that works well. "You don't kill the cow that feeds you," said one manager, alluding to potential political intrusions into the management of the industry. But government economic worries have already affected the Saudi NOC. First, the Gas Initiative was widely regarded as a political initiative to attract foreign capital to the country at a time of low oil prices. The injection of foreign capital into the gas sector was meant to free more capital for Saudi Aramco to invest elsewhere. Second, several managers felt that the challenge of finding jobs for the growing population was the main threat to the healthy relationship Saudi Aramco has enjoyed with the government so far. An upstream manager asked, "Are we going towards becoming a welfare company which has to give employment to the people or should we strive to reduce employment to reduce our operating running costs? Employment is the company's highest net direct expenditure. . . . There is pressure from government asking why we can't do something, or how is it even possible that this guy who has a good national degree (among the best in his class) can't get a job in Saudi Aramco? This will become harder [with increasing government needs]. We can't solve the problem of employment. With the population increasing, we won't have enough jobs. Therefore we should focus on growing income for the government, not on providing jobs.

Trying to reduce unemployment is a losing battle. But we can make a contribution to diversifying the government's income."

In Iran too, a number of NIOC managers felt that the state's budgetary constraints would have an impact on the industry. As a company executive put it, "If government gives us enough capital, we can produce more. If the government pressures us, we won't be able to operate." This problem was recognized even at the Ministry of Petroleum, where a key official explained that NIOC's revenues were reduced every time the ministry or the government needed extra revenue. A senior executive discussed the company's ambitions to emancipate itself from this stranglehold: "Yes, this is a real problem. . . . There's more profit with a commercial NOC. But the government has a short life. . . . The dependence on oil revenues for the government's budget is an important factor to determine whether it's a good time to allow the NOC to be very commercial and very independent or not."

It appears, therefore, that the fate of the transition of NIOC to more commercial operations hinges on the government's budgetary situation. Under the proposed reform of state–NIOC relations being discussed in the Majlis, it appears certain that the government will maintain the same returns it currently gets from NIOC (see the section below on "Financial Structure"). However, its great dependence on the oil rent means that the government is reluctant to take any risks by letting NIOC manage itself. High oil prices and the diversification of the economy should, in theory, encourage it to give the company more autonomy.

In all the five cases investigated in this book, NOC managers appeared to be quite aware of a link between the government's dependence on oil revenues and their lack of autonomy of operations. Numerous individuals highlighted the importance of diversifying the economy. It was clear that greater economic activity in other sectors meant more jobs and more wealth that government would not be seeking from the NOC. Diversification should, of course, be a prerogative of government, and some I spoke to in Saudi Arabia, Iran and Algeria expressed frustration at the government's inability to act effectively in this matter.

Evolution of NOC Missions

Since their creation, national oil companies have assisted governments not only in developing the national economy through the transfer of oil export revenues but also with a number of programs designed to provide services to the population and to promote local capacity. In many cases, the scope and objectives of the NOC's mission were not the object of clear policy-making by government or companies. After the oil crisis of 1973, there was a rush to build a welfare state, and spending programs were reactive to oil prices and erratic. But government expectations of the scope of the companies' missions are changing, and within the NOCs there is a constant struggle to find the right balance between the national and commercial missions.

Past Trends

In the early days of the petroleum industry in Iran, Kuwait, Saudi Arabia, Abu Dhabi and Algeria, the foreign oil companies had to make up for deficiencies in the public infrastructure in order to carry out their operations effectively. This became part of the NOCs' larger national mission when they took over. They carried out a variety of services that could be purchased from the private sector in mature market economies or procured from other state agencies in planned economies. In most OPEC countries, NOCs have operated airports, communication systems, roads, social clubs, hospitals, housing, universities, and so on. They became instruments of the public treasury and their export revenues have provided collateral for the public debt. They have also supported the balance of payments, promoted industrial, commercial and infrastructure projects and employed and trained nationals. In addition, they have been instruments of foreign policy. Over time, states came to rely on them. Some NOCs became states within the state. Some grew beyond the control of government, which was handicapped by a growing knowledge and capacity gap vis-à-vis the NOC. This enlargement of their mission led the managers of NOCs to become preoccupied with sensitive political questions, and it distracted them from the priority of managing the oil and gas industries (Boussena in Chevalier, 1994: 34). Of course, the management of this extensive national mission also came at a significant financial cost for these companies, which was particularly evident when the profits from high prices did not mask it.

Finding the appropriate balance between the NOCs' commercial and national missions (that is, between their core and non-core missions) depends on the state of development of the non-oil sector of the economy and the government's capacity to deliver those services itself. If the non-oil sector cannot deliver supplies and services to the high standards to which the NOC operates, there is often very little alternative but for the NOC to undertake these operations itself. As for the state, it took some time before the relevant ministries took on these responsibilities and gained the necessary insti-

Figure 6-1. Changing Missions of the National Oil Companies

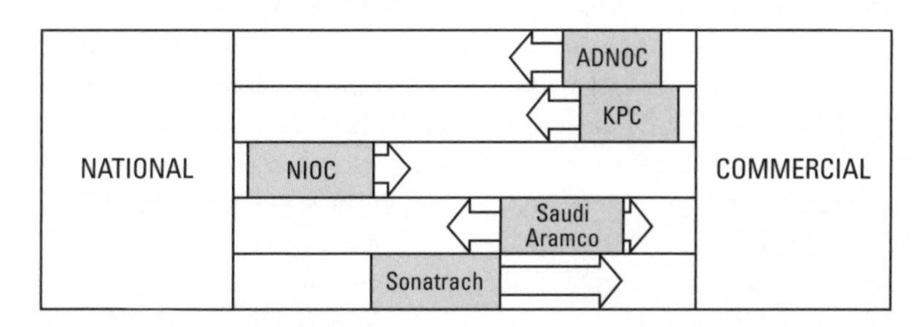

tutional capacity. The devolution of this national mission to the state and the private sector was prompted by growing state institutions, constrained oil revenues, a growing private-sector capacity and the increased commercial ambitions of NOCs.

Figure 6-1 shows broadly how NOC managers and oil ministry officials perceived the NOC's mission to be changing. Sonatrach and NIOC are now expected to behave more like commercial entities; Saudi Aramco is directing greater effort toward promoting the national economy while keeping a strong focus on its bottom line. ADNOC's and KPC's missions are relatively stable, though they face greater pressure to employ locally. The Abu Dhabi government has also asked ADNOC to prepare the national workforce for employment through training.

Less Need

The state is now stronger in these NOCs' countries and is able to set up and administer health programs, build infrastructure and provide essential services to society. NOCs are no longer asked to spend on infrastructure development. In Iran, for instance, NIOC was responsible for "national mission" expenditure some 30 years ago when the state infrastructure was weaker, but this is no longer the case. It now spends on infrastructure development only in the areas where it is producing and as required by industry needs, as does the oil and gas industry in Iran as a whole. This spending is channeled through regional authorities, which have put in place the necessary programs to serve the population. Many industry managers in Iran felt that it was crucial for governance and for avoiding social conflict that national mission expenditure should be transferred to regional authorities rather than be spent directly by companies. This transfer of responsibility signals the government's jealous control of spending prerogatives and also of the state's increasing maturity. And NIOC's disengagement from national mission expenditure is in fact more a function of the state's growing capacity than of the commercialization of NIOC, which remains in many ways an instrument of the state. A ministerial adviser's comment highlights the changes in the state's capacity: "Iran has that capacity now. So politicians don't expect the oil and gas industry to solve the problem of employment. The politicians are focusing on the issue of receiving oil revenues and how to spend them."

In some cases, NOC employees regret the devolution of national mission activities to the state, as they are aware of degraded standards. The NOC usually has access to superior technology and management processes, the know-how to manage projects, and has often committed large funds to these programs. In Saudi Arabia, for example, interviews indicated that there was some concern about the state's capacity to take on Saudi Aramco's social programs. The following is an excerpt from a group conversation at the company:

> VM: "Shouldn't some of [the NOC's] community functions be the responsibility of government?"
>
> SA25: "The government is learning from Saudi Aramco."
>
> SA21: "I don't agree. The government took over the schools from Saudi Aramco and didn't maintain the standards."[1]
>
> SA22: "The government agencies are learning from our practices."
>
> SA23: "The government relies too much on us. The money comes from the company and then the government should develop the country."
>
> SA20: "I don't see the effects of Aramco on our brothers and sisters outside the company. We are selfish. Where does this leave everyone else? We have the best treatment, the best salaries, the best hospitals . . . We worry that everything will collapse if the company lets go."

Within the oil ministries, however, it was not clear that NOCs added value with their contribution to the national mission. In cases where the national oil company provides a service to the highest standard, comments made indicated that government agencies would be able to do so at a lower cost.

1. According to written comments made by the company, this is not strictly accurate. Saudi Aramco built 135 schools in Eastern Province for the Saudi government but it never ran them.

Because of the costs associated with their peripheral spending (in activities such as building hospitals or hiring extra staff), NOCs are often deemed to be inefficient.[2] However, it is not national mission spending in itself that makes an NOC inefficient; it is poor organizational processes guiding the execution of its duties (whether of a commercial or national nature). An NOC is not inefficient if it spends money on programs of national interest in accordance with government directives; it is inefficient if this money is spent haphazardly, without strategic guidance, or without concern for measuring the success of the expenditure. This applies to all spending, whether for a training program for young nationals or for upgrading a refinery. In other words, the efficiency and performance of an NOC must be evaluated against its mission as directed by government. To the extent that its mission takes it outside the hydrocarbon sector, it is important to understand whether the business and microeconomic environment in the rest of the economy will serve to enhance the NOC's contribution or to dissipate it.

New Trends

There have been clear transformations in the national role of NOCs. Most governments no longer ask their NOC to invest in programs such as roads and hospitals that the welfare state can provide, but they do ask it to be a motor of economic growth. Promoting economic prosperity is no longer simply a question of maximizing oil rent; it is based on increasingly sophisticated strategies for stimulating the diversification of the economy. But it is not clear that governments have evaluated carefully how they can put to best use the NOCs' contribution.

For example, there is a greater emphasis in Kuwait, Abu Dhabi and Saudi Arabia on providing employment for nationals. NOCs in the Arab Gulf are asked by government to nationalize their workforce and to hire more nationals. This is perhaps the first area of government intrusion into the management of the oil industry at present. NOC managements are sensitive to the needs of the population, and interviews indicated that few were opposed to this requirement. Some mid-level Saudi Aramco managers wanted to see even greater company efforts in this direction. A geologist explained: "I pressured the VP to hire Saudi physicists. I said: 'I will train them. I will put in the work to get Saudis trained for the job as well as expats.' This is a[n] NOC!" In

2. For instance, see Charles Rivers Associates (2002) and various articles in *Economies et Sociétés*, Série Economie de l'énergie, "L'avenir des sociétés nationales des pays exportateurs d'hydrocarbures," EN, no. 6, September 1994.

Kuwait, however, many managers objected to what is perceived as an exaggerated government emphasis on employment numbers rather than quality. The following comment illustrates their concern: "Forty to fifty Kuwaitis are suddenly incorporated into marketing in one year. How do you incorporate so many people [without experience]?"

In Algeria and Iran, the national oil companies have already nationalized their personnel, and they are no longer encouraged to hire more nationals. In these countries, the state's expectations regarding employment opportunities in the NOC have changed over the past 30 years. This change is presumably a result of its realization that the oil industry is not labor-intensive and that the populace would benefit more from an increased transfer of oil rent to the state than from the creation of several thousand jobs. A refinery in Iran, as a case in point, has received directives to reduce the workforce from 1,200 to 600 in order to increase its efficiency. In spite of this change in government directives regarding employment, it is often difficult for most NOCs to justify layoffs. Punitive management practices are contrary to the corporate culture, and in the case of Sonatrach, they are opposed by the company's strong union. But Sonatrach may have to re-examine its employment policy now that the hydrocarbon reform bill has been approved by parliament because the introduction of competition may force it to streamline its operations. The company's management, as well as government, has argued that reform will bring more jobs and that the hydrocarbon law gives oil companies incentives to invest in training for new skills.

The gradual diminution of NOCs' responsibilities in the social sphere was also a result of the sudden contraction of revenues when the price of oil crashed in 1997–98. This slump prompted Saudi Aramco, like a number of other national oil companies, to cut its national mission expenditure. But, unlike the situation for NIOC and Sonatrach, economic challenges to the stability of its operating environment have since renewed Saudi Aramco's interest in supporting the national economy. In doing so, it appears to be preempting government pressure. Its thinking about investments is innovative: it seeks what it calls the "Golden Quadrant," where investment can be profitable, provide a public service and satisfy international market obligations. This approach is interesting because it views commercial and national functions as complementary rather than opposed: the company can contribute to the economy in ways that are consistent with its commercial mission. For instance, Saudi Aramco will provide training for nationals in chosen sectors in order to increase the pool of skilled workers available in the future. This approach requires patience because its dividends are longer-term ones.

However, there are limits to its responsibilities to the kingdom. The message in interviews was consistent: Saudi Aramco would not return to infrastructure-style spending. This time, the company is turning work over to the private sector. In essence, it is increasingly outsourcing in order to expand local capacity and the country's productive forces. The rationale is that these productive activities will contribute to the diversification of the economy and, in turn, to lessening the dependence of the non-oil economy on Saudi Aramco's generation of oil export revenues. But before handing over the work, Saudi Aramco is helping to create capacity in the private sector. As an Aramcon explained, "There's been a change in policy. We focus on our core business. We have been building roads and schools with the companies to train them to the highest standard and now we will hire them to do it alone."

In Algeria, the dwindling base of oil reserves has added urgency to the need to curtail Sonatrach's spending as it develops the remaining reserves. The Ministry of Energy and Mines has encouraged it to cut back its social spending and focus on its core business. The substance of the company's national mission is thus changing as a result of being driven by the company needs. In a way of thinking that appears similar to Saudi Aramco's, a director responsible for Sonatrach's civic engagement explained that the company will maintain its "deeply anchored tradition" of investing in, for example, training but that this activity will be more organized and focused on Sonatrach's needs. He explained further that "[this activity] used to be . . . a result of demands made by government. Now it is a policy of the company. It is something we want to do for our image. We want to be a civic company (*une entreprise citoyenne*)." The company's social works department is active in healthcare, sponsors football clubs and responds to emergency needs of the population (such as those resulting from the 2003 earthquake and the flooding in Bab El Oued in 2001).

Here is where Sonatrach's rationale differs from that of Saudi Aramco. It is turning to direct spending on social programs, as opposed to deeper spending for long-term socioeconomic benefits. Its new social programs are more like the corporate responsibility programs of IOCs, which aim to foster a positive image of the company—what an outspoken Iranian manager dismissively called "charity"—but it is nonetheless a new trend in NOC spending. Deeper spending corresponds to the historical national mission of NOCs that often bears fruit only years later. This social investment with a long-term view has been a characteristic luxury of the national oil companies, one that is being challenged in several countries by the imperatives of the bottom line and containing costs.

Sonatrach will increasingly disengage from heavy spending on its national mission, and this will make it more commercially focused. But such change is slow to take effect. Its involvement stretches out across the territory, a result of heavy investment over decades. A human resources director mused, "Are we a state within the state? We are certainly a national asset; and for all we contribute, we cannot be privatized."

The company is in the midst of a major transition resulting from the new law for reform of the hydrocarbon sector (see above). This law sets out clearly the terms for introducing competition to the sector and enables the commercialization of Sonatrach, but not its privatization. Its new mission is clearly more commercial, but how far will the company go in this direction? Is it to become focused only on its core business, like any other oil company operating in Algeria, or will it remain a public service company? The transition to more commercial thinking is contrary to the established "culture of the heart" in Sonatrach as well as to societal expectations. For this reason, it may be reluctant to give up its national role. Sonatrach has made itself indispensable to Algeria, and the company draws a good measure of its pride from this social generosity. Employees seem to feel that the national mission (at least in its lighter, short-term "charity" form) should be maintained and that layoffs should be avoided. However, as noted, they also want Sonatrach to compete nationally and internationally with the best private companies. The coming years will show how far Sonatrach will need to focus on its core business in order to accomplish this. In any case, the transition in corporate identity and culture is likely to take some time.

Financial Structure

Understanding how NOCs generate, retain and spend revenue is essential to understanding how they may be affected by future government fiscal needs. For instance, if an NOC works like a government ministry and requests its funds through budget allocations, government will be able to control its spending—so much so that it could sacrifice the needs of the industry to its own budgetary emergencies. To insulate themselves from the volatile budgetary needs of government, NOCs should be able to accumulate reserves, invest according to a government-approved plan and work within a fiscal framework that enables long-term commitments to capital expenditure in the hydrocarbon sector. These commitments should be at least as stable as those made by private-sector companies of similar size and profitability. Where the NOC does not have sufficient capital, its management may seek other means of financing, in activities that may escape the day-to-day control of the state.

Table 6-1 compares the financial characteristics of the relations between the state and the five NOCs under study here.

Saudi Aramco has inherited the fiscal structure that existed under the concession system, when it was under foreign ownership. The company pays royalties and dividends to the state, which now amount to approximately 93 percent of its profits. The remainder never leaves the company, which therefore "does not have to ask permission from the government to get the money back." Saudi Aramco enjoys comfortable financial operating conditions. From interviews, I gathered that the company has sufficient capital to self-finance investments in new capacity. As we saw in chapter 4, the decisionmaking process for budgetary and strategic plans also protects it from the everyday priorities of government. This is essentially because the company's representation in the Supreme Petroleum Council is so "high level" that this protects it against interference from other state actors, such as the Ministry of Finance.

Saudi Aramco's financial arrangements encourage it to maximize profits for the government because it also maximizes its own returns in doing so. However, this mechanism does not impel the NOC to cut costs or to set up lean operations, because it has money to spare. On the other hand, it was apparent in interviews that one of the company's greatest incentives to operate commercially and to maximize profits is its desire to maintain the government's trust. In this context, its operations probably will become leaner when the government signals that its budgetary needs have become more pressing.

Similarly, ADNOC pays royalties to the Abu Dhabi government and then retains the funds necessary to cover its operational needs and capital costs and to maintain a "buffer" (similar to KPC's "legal reserve") for covering unexpected costs. ADNOC is taxed on this net amount by the Abu Dhabi Department of Finance, and the remainder is sent to the Abu Dhabi Investment Authority, which invests in all sectors for the UAE government. Interviews in ADNOC indicated that the company is satisfied with the existing fiscal arrangements. Managers often said of the company's cut that "We take only what we need." So far, it appears that the government agrees.

Sonatrach too enjoys clear terms regarding its fiscal arrangements with the state. These have been governed over the past decade by the Hydrocarbons Law of 1986 with 1991 amendments. However, the new Hydrocarbons Law (No. 05-07 of 19 April 2005), which was due to come into effect by the end of 2005, sets new fiscal terms. Both fiscal regimes encourage Sonatrach

Table 6-1. Characteristics of State–NOC Financial Relations

Financial characteristics	*ADNOC*	*KPC*	*NIOC*	*Saudi Aramco*	*Sonatrach*
Government take	Government takes royalty and taxes. Surplus revenues are sent to the finance department of Abu Dhabi and the remainder to the Abu Dhabi Investment Authority.	Government takes revenues from crude oil sales. 10% is automatically allocated to the Reserve Fund for Future Generations. Government's share of revenues amounts to 80–90%.	Government takes revenues from crude oil export sales (paid to the treasury). It levies a special tax when required. Surplus revenues are sent to the Oil Stabilization Fund.	Government takes royalty and dividends amounting to approximately 93% of Saudi Aramco's profits.	Government takes royalties on the NOC's gross development revenues and between 30% and 70% of direct income tax. Surplus revenues are sent to the Fonds de régulation des recettes—a stabilization fund.
Company take	ADNOC retains funds for operaional and capital expenditure before taxation.	KPC buys crude oil and gas from the government, with a discount for its operating costs and a 10% legal reserve (as security against future claims). It sells crude oil and gas for the government and gets a marketing fee of about $0.50/b.	NIOC takes all revenues from product and gas sales (export and domestic), sales of crude for domestic use, as well as revenues from sales of government's share of the buybacks' crude allocation. It enjoys free use of crude for feedstock.	Saudi Aramco retains approximately 7% of its profits.	Sonatrach retains a share of product and crude oil sales and of its net returns on investments.

Available capital	Yes.	Yes. Funds are accumulated but margin on upstream sales is low.	No. Average annual investments are $3 billion. Company/state has borrowed from capital markets.	Yes.	Yes. Company has borrowed from capital markets.
Transparency	Internal transparency.	Internal and some external transparency. Annual report with financial disclosure, but data are aggregated.	No. Opaque accounting processes.	Internal transparency. No financial disclosure outside the Supreme Petroleum Council.	Internal and external transparency. Annual report with detailed financial disclosure is available online, but some data are aggregated.
Reform prospects	None.	KPC is negotiating for a share of the upstream sales.	Reform bill. NIOC pays a royalty of 50% and a minimum of 27% percent in direct income tax.	None.	Further steps are being taken to improve the transparency of accounting practices.

Source: Author's interviews.

and foreign oil companies via tax reductions to spend more on finding and developing new reserves in Algeria. With the implementation of the new law, the foreign oil companies and Sonatrach will each pay royalties on a sliding scale related to volume produced. Royalty payments will also vary with the maturity of an area's development. Differentials of up to 7 percent in royalty rates will apply. The petroleum revenues tax based on profit will start at 30 percent and rise to 70 percent. From the foreign oil companies' point of view, these rates are an improvement from the previous levels (a royalty of 10 percent on production and a profit tax of 42 percent) because the sliding scale system offers greater incentives for investment.[3]

In Iran, the government takes all revenue from crude oil export sales. NIOC receives crude for internal use free of charge. Before the third five-year plan, NIOC received a budget allocation from the government, like any government institution. But now it can generate its own revenues. These come from gas sales (including gas products), oil products that are exported,[4] crude oil sales for domestic use and assets created through the government's stake in buyback contracts. The revenues from the sale of buyback crude belong to NIOC and pay for its current costs. The company does not pay taxes to the government on these revenues if they are reinvested. But when the government is short of cash, it levies a special tax on government agencies, and NIOC must draw on its revenues. According to its management, it is allowed to invest an average of $3 billion of its revenue per year.

NIOC is on a tight lead when it comes to finance. Its budget is prepared by the Ministry of Petroleum and must be approved by the Ministry of Planning and then by parliament. The industry is integrated in the government's financial system, and as a result, the capital needs of the hydrocarbon sector are frequently sacrificed to those of other ministries. A senior company executive explained that the industry does not dispose of sufficient capital to achieve the targets set by the government's annual development plans. "We have nearly sufficient revenue to accomplish the third five-year plan. But for the fourth five-year plan, we need more." The hydrocarbon sector has therefore turned to foreign investment to provide the shortfall in capital.

The (current) seventh parliament has reviewed a bill for the reform of NIOC–state relations. This includes the controversial removal of the energy subsidy system and proposes a new structure of relations between the state

3. *http://www.mem-algeria.org/fr/legis/evo_hyd_legal_r.pps*. The legislation is at *http://www.mem-algeria.org/fr/legis/*.

4. Domestic oil products are sold at a subsidized price and therefore do not provide revenue.

and the NOC.[5] In terms of payments to the state, the reform will lead to the introduction of a tax and royalty system for NIOC. Transfers to the state will include a royalty of 50 percent of gross crude oil sales and a royalty of $.01 for every cubic meter of gas.[6] A special income tax for oil of at least 27 percent will be levied on revenues after the exploration for and the development and selling of the crude. Revenues from the special tax will go to a government account and can be used to pay subsidies for gasoline or other government programs. Ownership of the natural resources remains with the government of the Islamic Republic of Iran, but any oil produced from NIOC's activities belongs to it after payment of royalties and special taxes.

In the proposed reform, the royalty to be paid is unusually high, 50 percent. The largest portion paid to government is normally the tax on profits (royalties are paid before profit). However, according to senior figures in NIOC, the government, in particular the Guardian Council, was worried that the "government take" could be less with the reformed taxation system. Indeed, the Guardian Council objected to the provisions for the payment of royalties. From interviews, I gathered that the new fiscal arrangement would probably not make more money available to NIOC but that it would provide crucial space for the company to operate more commercially. Its managers appear to be less concerned about revenues than with the clarification of its prerogatives. As one senior executive explained, "We don't want more revenues. We want more clarity in our relations with the state and more commercial behavior. We want to be a commercial entity."

The Ministry of Petroleum jealously controls NIOC, which is to some extent the focal point of power struggles between the principal state institutions of Iran. There are indeed many voices speaking on and for the sector and many hands trying to acquire its revenues. One ministerial adviser explained that an important challenge for restructuring the fiscal regime was the Ministry of Petroleum's right to dispose of a share of NIOC funds in order to pay its employees better salaries. But if the relations between NIOC and the government are clarified and made more transparent, these funds

5. The sixth parliament passed the bill just before the end of its mandate, but then the Guardian Council blocked it. The bill was sent to the seventh parliament for a review of specific elements of it. These were still being discussed in late 2005.

6. The royalty for what Iranians called "unsure fields"—that is, those shared with neighbors—is 80 percent. The royalty for offshore fields is 90 percent. Gas injection for the management of the oil fields is not subject to royalty. But there will be a royalty for flared gas starting in 2009. The price of oil developed and sent to domestic refineries is based on 95 percent of the export price for every month. This does not include the cost of transportation and putting it in storage tanks, which accounts for the extra 5 percent.

will no longer be available to it. This has been an obstacle to restructuring the relationship between the ministry and NIOC. President Ahmadinejad has expressed his intention to carefully examine the financial relationship between government and NIOC, but it is unclear at the time of writing (late 2005) what impact this will have on the prospects for reform.

In Kuwait, 80–90 percent of oil revenues go to the state. The Kuwait Petroleum Corporation purchases the crude oil and the natural gas from the government after production. It deducts from the purchase price its operating costs (people, salaries) and 10 percent for a legal reserve,[7] as allowed by the Supreme Petroleum Council. Capital expenditure is not deducted—this comes out of KPC's budget. Then it sells a share of the crude and gas on the markets at a price set by the oil ministry and returns sales revenues to the state.[8] It does not get a profit margin on the upstream sales but it takes a fee of about 50 cents on the barrel for marketing the crude oil. Upon purchasing the crude and gas, KPC can also refine them and sell the products.

This financial structure explains to a large extent the concern the company has about the productivity and profitability of its ancillary operations, such as international refining. KPC earns its revenues from fees for marketing the crude, from returns on funds invested and from returns on the transformation of the resources, and the government foots the bill for upstream operating costs. The upstream operating company, the Kuwait Oil Company, works as a cost center: it is a division of KPC that adds to the costs of the business but only indirectly contributes to its profits. Other subsidiaries run profits, by refining the crude produced by KOC, for example. Those subsidiaries are profit centers, and they function as businesses in their own right; their managements are responsible for expenditures and revenues. In this system, KOC is meant to be free to focus on its core upstream function. But there is a lack of concern about containing upstream costs at KOC and KPC: because the government pays for the operating costs, the parent company has little incentive to run KOC efficiently. There have been discussions in Kuwait about a possible reform of KOC's status as a cost center.

Nonetheless, according to a KPC finance director, the present system does provide the company with sufficient capital to reinvest in the oil sector annually. The availability of capital is in part attributable to its careful accumulation of funds and investments. The main challenge for KPC is how to spend

7. The legal reserve is the sum of money that a firm is required by law to set aside as security. This is designed to ensure that companies will be able to meet future claims and obligations.

8. Ten percent is automatically allocated to the Reserve Fund for Future Generations and is not available for current expenditure.

this money, as its uses are carefully monitored by the tendering and auditing committees of the government and the company.

Among the financial arrangements outlined above, the royalty, tax and dividend systems are better able to insulate the company's budget from government needs. That is because the mechanisms for the transfer of the oil rent are clear and institutionalized, as opposed to systems in which the NOC is akin to a government ministry and its budget is subject to changing government priorities. NOCs require a clear and predictable system for retaining or generating revenues. They operate in a commercial environment that demands long-term capital commitment. The financial systems in which ADNOC, Saudi Aramco and Sonatrach operate have the advantage of letting the NOC retain a share of upstream profits. The competing model, which gives government greater control over the flows of revenue, has all profits (notably from crude oil sales) go directly to the treasury and requires the NOC to request funds like any other state institution. The Kuwaiti and Iranian NOCs' financial systems exhibit some aspects of this type of arrangement, notably because crude oil sales go directly to the treasury (less their costs), but they have other sources of direct income of which they can dispose, subject to budgetary approval.

Availability of Capital

If the NOCs' access to capital for reinvestment in the hydrocarbon sector were constrained in the future, could they readily turn to the capital markets for project finance? Interviews were conducted with four Middle Eastern banks in order to assess the availability of capital and the banks' impressions of NOCs' future investment needs. These banks indicated that they were keen to finance oil and gas projects and would make capital available when asked to do so. Islamic bankers explained that there was $200–$250 billion in Islamic funds available. Other bankers thought conventional banks would play a larger role in funding regional oil and gas projects. One Kuwaiti banker added, "We don't believe in Islamic banking. It's hubbly bubbly."

On the whole, the bankers felt that even though NOCs had long had "money lying around" and were still able to self-finance their projects, they would soon turn to the capital markets. According to one banker, "In the past, Kuwaiti and Saudi energy projects were state funded (with the exception of petrochemicals). This will change. Qatar has gone to the market for finance. The Kuwaiti and Saudi governments have been running deficits because the projects have been state funded (including the North fields project)."

This banker's assessment was that NOCs will privatize some of their activities in order to generate capital and to lessen the financial burden on the state. This change will affect primarily downstream activities, such as petrochemicals, refining and power. The NOCs' obligations are regarded as part of the state's obligations, and may be secured, as in Mexico, by offshore affiliates that incur the debt and sell the oil that generates the revenue to service the debt. However, this option is not available to Iran and Kuwait, where revenue accrues directly to the government instead of the NOC.

Sonatrach has had more experience of the financial markets than the other NOCs. Even though it no longer receives a state guarantee for its loans, the company is apparently "not finding it difficult to borrow"—in fact, the cost of capital has decreased since Sonatrach's finances have no longer been strictly bound to those of the Algerian state. ADNOC and NIOC have also turned to capital markets in the past. But Saudi Aramco and KPC (and its subsidiaries) do not generally know how to approach capital markets for project financing. Saudi Aramco has gone to the markets only once, for a loan to buy its fleet of new tankers. KPC's tanking subsidiary has so far financed its operations from its own equity, according to a regional banker; but it will seek finance on the open market for its new tankers. It will do so in order to establish the process and to get to know how to borrow, not so much because it needs the money. A regional banker explained that when NOCs decide to borrow, "they don't know where to go because they don't know the financial markets and don't know how to borrow."

Good Financial Governance

The government should naturally be concerned with the good governance of the revenues that are left with the NOC. Optimal financial systems will maximize the rent available to the state and clarify the costs of the NOC's activities. Among the financial systems described above, the one most likely to successfully drive the financial performance of the NOC appears to be Sonatrach's, in which operating costs are taken out of the company's available capital. In the cases of ADNOC and (as it appears from more limited data) Saudi Aramco, the companies pay their operating costs after receipt of royalties but before sending dividends back to the state. KPC also deducts its upstream operational costs from the purchase price of crude oil from the government. NIOC internalizes its operational costs (described below). All these mechanisms create opportunities for the NOC to absorb rent.

KPC's and Sonatrach's financial systems encourage them to be self-reliant in terms of financing because they own facilities and can generate revenues from their investments. They supplement their funds through independent, essentially international, investments. In NIOC's case, it has turned to independent sources of capital, such as loans and the sale (and trading) of its share of crude oil from the buyback contracts. This trading is handled by the Naftiran Intertrade Company (NICO), an offshore company registered in Jersey, Channel Islands, and entirely owned by the National Iranian Oil Company. NICO pays no taxes to Iran. It is designed as a way for NIOC to raise capital. Reliance on independent sources of revenue (from international refining or trading, for instance) drives KPC and NIOC to conduct their operations in these sectors efficiently. However, their exploration, production and marketing of crude may suffer from inefficiencies because they receive no direct financial gain from these activities.

Efficient and clear internal accounting processes are a further critical component of good financial governance. When different subsidiaries participate in the transformation of crude oil or natural gas into refined products, the transfers of feedstock and funds between the parent company and the subsidiaries must be transparent so as to minimize waste and encourage profitability. In Iran, the mechanisms for these transfers have, until recently, given subsidiaries few incentives for cost cutting. Under the current procedure, which is being reformed, the refining companies prepare budgets with requests for the necessary volumes of crude (and gas or wastewater) but with no consideration given to the market value of the feed. The refining company NIORDC pays NIOC for these feeds at an internal price. I was surprised to learn from a manager at NIORDC how they arrive at the internal price:

"We have a system: we add the current budget and the capital budget for the company, and this total is the spending that we divide by the number of barrels. Say, for example, we get 100 barrels from NIOC and our total spend is \$200, then 200 ÷ 100 = \$2 per barrel. Therefore, we give this price to NIOC as our internal price for crude."[9]

The same procedure applies to the Iranian refining companies, which are compensated for their capital and current spending less the cost of the feed to NIORDC—in other words, their profits and losses are sent back up the organization. In 2004–05, however, new measures have been introduced to decentralize the structure of NIOC and to increase the transparency of financial flows and make the subsidiaries and companies more accountable for their spending. The companies that make up NIOC (as well as the gas company

9. It is noteworthy that a retired oil executive recognized this pricing system as the one the international oil companies used for managing their budgets in exporting countries before 1973.

NIGC and the petrochemical company NPC) now pay taxes to the government. Every subsidiary and company has its own board and can now make its own purchases directly. Soon, refining companies will buy the crude (and other feed) at market prices and sell the products at market prices. If the subsidiaries have shortfalls in their budgets, they will have to turn to the capital markets. As a manager on the ground put it, "If there's a problem, we are on our own." Generally, the performance of the subsidiaries will become more apparent in this system. The management at a refinery visited in Iran was surprisingly supportive of the coming challenge of accountability.

In Abu Dhabi and Kuwait, some costs are also absorbed through the company's network of subsidiaries. However, the greater transparency of these internal costs allows a more critical examination of them. ADNOC, like KPC, is a parent company overseeing various companies, some of which operate like cost centers. For example, the Abu Dhabi Company for Onshore Oil Operations (ADCO), which explores and develops oil, does not generate revenues (because ADNOC markets the crude) but requires a $1.5 billion budget per year. Other companies, such as the National Drilling Company (NDC) and ADGAS, send their profits to ADNOC. Different companies have different fiscal arrangements, and some receive gas for feedstock at a rebate. Abu Dhabi Gas Industries Ltd. (GASCO), which is a sole-risk company, pays an internal price for gas that is accounted for in its annual report (for internal use only). In Kuwait, internal accounting processes put pressure on profit centers' subsidiaries to show profit.

Transparency is increasingly seen as a means to optimize the financial governance of the hydrocarbon sector. Rigorous accounting procedures are gaining ground in the five NOCs under study. But this drive for improved corporate governance is not necessarily aimed at increasing public disclosure of information. Sonatrach is the only NOC studied that presents a public annual report with a breakdown of costs and transfers. And this information remains imprecise because the data are aggregated, but it is a vast improvement over past Algerian accounting standards, which could be understood only by the Algerian state. This type of reporting is not very different from that of the private oil companies. Sonatrach is now preparing the transition of its accounting practices to U.S. Generally Accepted Accounting Principles. In written comments, the company explained that this had reduced the cost of borrowing on international capital markets. Its processes are changing partly in response to increased contact with the international environment, including working with foreign partners, but also due to the influence of the World Trade Organization and to Algeria's partnership with Europe.

Preparation for accession to the WTO is also having a positive (though lesser) effect on Iran and Saudi Arabia. A Saudi oil ministry official explained that the "WTO is not as intrusive as you might think." He explained that the new requirements have only meant that an independent, foreign auditor and an internal auditing committee check Saudi Aramco's financial accounts. The information collected goes to the Supreme Petroleum Council only. ADNOC's accounts are exposed to a similar auditing process behind closed doors.

Though these internal accounting practices are beneficial to good governance of the hydrocarbon sector, the lack of external transparency is an obstacle to access to capital markets. On this point, a regional banker felt that there is adequate capital available to finance the oil and gas developments in the region but that transparency would be a crucial issue for NOCs needing to raise capital. The information transmitted on the oil and gas sector is sketchy: "We get the big picture only. Things can be distorted easily. Is there patronage? How do we know? Cost overruns are not disclosed either."

A senior Aramco executive explained that the Saudi government does not allow it to divulge financial data. When the company went to the financial markets for its fleet of tankers in the mid-1990s, a Saudi banker explained, the company prepaid the loan in order to avoid divulging any further its financial information. Presumably, it did so to avoid embarrassing the government with disclosures of the rent it transfers to the "black box"—in other words, the company does not disclose the amounts that the government receives from the oil sector. I discussed this with a vice president at Saudi Aramco:

> VM: "The people have a right to know. There should be democratic oversight of your spending . . . and accountability for your management of the national resource."
>
> SA: "But what right do they have when they pay no taxes and when the government provides all the needed services?"

But an oil ministry official felt that this would change. There are elements in the government arguing for greater transparency. For example, members of the Shura, the consultative council, have reiterated their request to the king for greater disclosure on financial matters.

Conclusion

In the national oil companies, there is a pervasive concern about their countries' future economic troubles and the lack of economic opportunities outside the petroleum sector. NOC managements are aware that they cannot solve their economies' structural problems, and they perceive a risk of increased pressure from governments to give them more income. The NOCs are unequally protected institutionally and financially from government intrusions. In anticipation of future economic challenges, many contribute to their country's diversification effort, for example by outsourcing to the private sector and providing training for nationals. But the boundary of responsibilities between NOC and state is increasingly respected, because NOCs are now focusing more on optimizing their commercial activities while leaving deep social and infrastructural spending to the state.

seven
Industry Challenges

National oil companies face two types of industry challenges, domestic and international; they have some control over the former but little over the latter. At home, NOCs deal essentially with capacity and operational challenges; that is, with improving their management processes, increasing their access to technology, acquiring greater experience and securing capital. In this respect, each NOC is unique and faces different challenges. Externally, the companies face similar industry threats, which are also common to international oil companies. These pertain to maintaining market share, competition from other producers, establishing new markets, protecting future demand for hydrocarbons and responding to new environmental regulations in the consuming countries. However, these common challenges will affect each NOC differently because each has different strengths and weaknesses and has established itself in its own marketing niche.

Domestic Industry Challenges

From comments often heard in consuming countries, one might assume that the greatest industry challenges a national oil company faces are OPEC politics and competition from producers outside OPEC. But this assumption reflects the concerns of the consuming countries, where people are preoccupied above all with the supply of crude oil to markets. In fact, because NOCs operate largely within their borders, they tend to be more preoccupied with their national operating environment and with domestic industry issues such as enhancing resource development than with international industry issues such as competition from rival producers. Internal challenges affect the various producers differently.

Upstream Technical Challenges

Mature Reservoirs. NOC professionals and some ministry officials in Iran, Kuwait and Abu Dhabi discussed specific challenges to the upstream sector. For most reservoirs, NOCs know which technology to apply. However, the great test these producers face is to determine how best to develop aging reservoirs, which present new geological challenges. As an exploration and production (E&P) director in the Emirates put it, the upstream challenge for these producers is to identify, develop and apply the right technology for managing major projects and difficult reservoirs. They can buy the technology in the form of a service provided by a specialized company such as Schlumberger or Halliburton;[1] but for enhanced recovery, engineers at KOC and ADNOC explained that their companies need experience more than technology because it is a question of knowing which technology to use in a specific set of circumstances.

In Kuwait, for instance, oil production has relied so far on large reservoirs, such as Burgan, where the lower portion of the reservoir is produced with the help of a powerful natural water drive. But as production reaches into more complex parts of the reservoir, more sophisticated production engineering is needed in order to recover the remaining oil. KOC feels that it needs the help of the oil majors to control the depletion of Burgan and other maturing reservoirs where "it will be critical to manage the water injection."

In Iran, reservoir engineers explained that they had a low recovery factor (that is, the portion of the reserves that can ultimately be recovered in the development life of a field)—this factor was estimated by the International Energy Agency at 27 percent. To maintain recovery in complex geological structures requires a precisely controlled injection of gas, brought from other Iranian reservoirs, or CO_2. Another challenge is the higher average natural decline rate of Iran's mature fields—it is estimated at eight percent onshore, against Saudi Arabia's six percent.

E&P professionals in Abu Dhabi also discussed the technical challenges of developing "difficult" reservoirs, mature ones for example. They felt that new concession terms would be needed in order to bring a greater commitment

1. NOCs outsource seismic studies to service companies, for instance.

by foreign partners to apply their technology and know-how to those fields. The terms of new or renewed concessions are debated heatedly, and one issue on the table is the conditions for investment in enhanced recovery. An E&P director for ADNOC explained that it wants to convince its foreign partners that despite marginal financial gains, each dollar invested in long-term recovery is worthwhile because it benefits Abu Dhabi. The company appears confident that this message will be received favorably. However, IOC upstream managers have complained of a lack of incentives to invest in enhanced recovery in Abu Dhabi and also Iran. Their NOC counterparts appear to have little appreciation of foreign partners' frustration with the terms of investment. For instance, in Abu Dhabi, NOC managers felt that the IOCs were willing to invest in enhanced recovery for the good of Abu Dhabi. There was unwavering confidence in the foreign partners' willingness to contribute in this effort. Meanwhile, upstream managers in Iran felt sure that the reluctance of foreign investors was a result of the country's political risk, not of the terms of investment. These misaligned expectations regarding the terms of investment are examined further in Chapter 10, which discusses the reluctance of IOCs to work as service providers and that of the NOCs to lose control over the development of reservoirs.

In Saudi Arabia, the Ghawar field, the biggest in the world, is produced by "secondary" recovery (water flooding). Despite its maturing fields, most respondents in Saudi Arabia discussed the technical and investment challenges of enhanced recovery only in very vague terms. In fact, at Saudi Aramco, no upstream challenges were discussed (in fact, few challenges were discussed at all—except the country's socioeconomic and political ones). Aramcons consistently sent the same message that is found in their brochures: Saudi Aramco is a reliable supplier of energy. It is not within my purview to determine whether the absence of comments on industry challenges was due to the actual absence of challenges or to the successful internal dissemination of a corporate message.

Exploration. Another challenge for the producers is exploration. Indeed, as it becomes more difficult to replace the barrels produced from the mature reservoirs, they will need to turn to the exploration and development of new fields in order to maintain production levels. This is an important challenge for the Kuwaiti Oil Company. A KOC executive explained that the company is equipped to see to the maintenance of its fields but that it needs more money, technology and skills in order to expand its capacity. Additional capacity will come from heavy oil reservoirs, which present new geological challenges to the company. A Kuwaiti engineer explained that "As a result of the targets set, we need to get to 4 million b/d in 2020. So they look at our reserves and figure we can do that. But to recover that extra 2 million will be difficult. . . . The extra 2 million barrels will all come from difficult reservoirs."

In Algeria, energy ministry officials argued that Sonatrach was not able to meet the exploration challenge on its own and that, as a result, Algeria needed to invite foreign companies to explore new reserves. I had expected this to cause friction with Sonatrach, but was surprised to find the exploration challenge tackled head-on in interviews with its executives. A corporate planner explained that only 40 percent of the country has been explored and that 70–80 percent of discoveries had been made before nationalization. Clearly, he said, the company has not explored sufficiently. This is essentially because it lacked the resources to do so, as is also the case in Iran. Moreover, exploration and production are riskier activities than development because misdirected exploration efforts can be a costly mistake and decisions to develop new fields must be based on sound commercial and geological analysis. National oil companies are often loath to take risks with public funds and may not have the experience necessary to tackle areas with new, technically challenging conditions.

Throughout the 1970s and 1980s, the fear of "risking big" led Sonatrach to devote more resources to developing liquefied natural gas (LNG) exports, which brought more immediate revenues for the state. On the energy ministry's side, attempts were made to attract foreign investors to explore and develop new oil fields, but its long-standing concern with preserving the country's sovereignty meant that the investment law was not sufficiently attractive to foreign capital. The revised hydrocarbon law of 1991 introduced an important change because it opened existing fields to foreign investment. This attracted the IOCs because it gave them the possibility of guaranteed returns in known fields (the hesitance of private international oil companies until that point indicates that they too preferred the known, less risky areas). Incentives for riskier exploration ventures included concessions regarding the decisionmaking process, arbitrage dispositions and fiscal terms. They were much needed, as Algeria was engulfed in the civil war of the 1990s.

R&D. Though many in the national oil companies expressed interest in learning best practices from IOCs, they are nonetheless developing their own skills independently. To equip themselves better in terms of technology and skills for upstream exploration and production, ADNOC, Saudi Aramco and Sonatrach are pressing ahead with efforts to participate in the research and development of new technologies that will give them an edge in the industry. This is a clear sign of how much they have matured since the early days of nationalization.[2] In this vein, Saudi Aramco has circulated documents through its NOC industry networks to promote strengthening NOC core competencies in what its CEO, Abdallah Jum'ah, calls TPP: technological excellence, people and processes. At the meeting in 2003 of the National Oil Company Forum, a new annual roundtable for CEOs of the world's major NOCs, he urged them to invest in state-of-the-art technologies and to make the recruitment and development of "top-notch" people a core competency.[3]

In ADNOC, E&P experts want to work with partners who give them exposure to the best technology. Much like Saudi Aramco, they want to be involved in the development of technology so that they know how to apply it and can ensure ownership of it. Many managers at ADNOC and its subsidiaries felt that developing their human resources was key to this process because without the right qualified people the technology was useless. Sonatrach echoes these ambitions about the development of technology. Officials from the Algerian Ministry of Energy and Mines hope that by working alongside industry leaders, Sonatrach will be motivated to spend more on acquiring new technology.

Upstream Investment Capital

Across the region, capital needs are expected to be compelling. Investments will be needed to increase production capacity, as non-OPEC production growth rates are slowing down and most OPEC countries were producing near to or at capacity in 2003–05. Saudi Aramco, KPC and ADNOC can self-finance projects, as long as their investment capital remains insulated from short-term government budgetary needs. But other producers need investment capital when their capital needs are sacrificed to the demands of the government's budget or their means of revenue generation cannot meet investment requirements, as in Iran and Algeria respectively. In Iran, the oil industry's internal revenues are clearly insufficient to meet its investment needs. As a senior figure in the Iranian oil ministry put it, "In any case, we have no choice but to go to international markets and investors. We have had revenues from oil but we haven't used them for oil. We used them for other sectors which don't attract FDI. We must use the oil rent there."

Iran and Algeria can turn to banks or to IOCs for capital. Capital markets meet Sonatrach's investment needs. Iran, however, is finding it difficult to

2. All three countries, together with Iran, also have specialized institutes or universities that are engaged in research and education and have established partnerships with universities and institutes abroad.

3. Remarks at National Oil Company Forum, Stavanger, 15–17 October 2003.

meet its capital needs for upstream investments in oil and gas, which a senior official of the Ministry of Petroleum estimated at 5–6 percent of the country's GDP. A Ministry of Planning official said that lack of budgetary support is the only bottleneck the Iranian oil industry is facing in its expansion plans. NIOC has relied increasingly on international sources of capital, for example foreign investment through the buyback scheme, European and Japanese banks and export credit agencies. As a chief adviser at the Ministry of Petroleum explained, "I can't attract capital to raise the production capacity [and offset decline]. So I borrow. I continue to borrow, but it's not enough. Not enough to improve the position of Iran in relation to its reserves [and to the place we should have]."

Iran also faces political obstacles in gaining access to capital. American lobbying played a central role in delaying a Japanese consortium's participation in the development of the giant 3 to 6 billion-barrel onshore oil field Azadegan; and in 2004, former U.S. Secretary of Energy Spencer Abraham warned European investors not to invest in Iran. However, they persist—the Iranian oil authorities are engaging foreign (non-U.S.) investors as stakeholders in the stability of the country by negotiating new deals for oil and gas developments. They have pressured notably the main existing partners, Total, BP, Shell and ENI, to sign up to new developments. Ironically, Washington's saber rattling may be to the advantage of those IOCs, which could benefit from softer fiscal terms from the Iranian authorities, anxious to draw them in.

New capital does not come cheaply. A ministerial adviser was forceful in his argument that it was the responsibility of consuming countries to facilitate investment in Iran's oil and gas industry if they want the country to supply the world with more energy. They could help by persuading the United States to lift its sanctions, reducing political risk and encouraging foreign companies to invest in Iran. He articulated his country's dilemma: under sanctions, Iran cannot develop its hydrocarbon potential, and this inability to deliver its supply of much-needed energy to the world could lead the stronger consuming states to invade it in order to develop its huge reserves.

Raising capital through foreign direct investment brings several challenges. All five producers discussed in this book plan to develop spare (idle) capacity. But those who develop that spare capacity with the help of foreign private oil companies may encounter some opposition from their partners. IOCs may not agree with producers about how much of the new capacity to keep idle. In Algeria, oil professionals and ministry officials expressed confidence that OPEC quotas will be increased and that investors will not be left with idle capacity. Elsewhere, however, plans are under way to develop or increase spare capacity; and if Iran and Kuwait succeed in attracting foreign investment to carry out these plans, frustration might ensue on the IOC side. In Abu Dhabi this is already an issue. An ADNOC executive addressed it when discussing the timing of government directives to increase production capacity: "We are all working to increase capacity, but the shareholders don't like idle capacity. We have a moral obligation to have spare capacity to supply the consumers . . . but now it's too late with the high demand. When the prices are low, there's no incentive to invest because you can't shut down capacity easily. (Look at Venezuela and the damage it did shutting down.)"

Because national oil companies do not have an exclusively commercial purpose, they are better able to carry out the oil policies of very large producing countries. Indeed, the ADNOC executive pointed to the difficulty, when foreign companies are involved, of developing spare capacity so as to balance markets. In Algeria, the tone was different, and energy ministry officials were adamant that foreign oil companies had not been deterred or even been concerned about the limitations on development as a result of OPEC quotas. "I participated in the four last bidding processes, and there was not one company that asked about our quota limitations in this process," commented one official.

The possibility of developing idle capacity is not the principal difficulty for foreign investors in Iran and Kuwait, however. The greatest difficulty there is political resistance to foreign private capital investment in the hydrocarbon sector. Opposition to foreign investment has emerged in parliament, the media and, in the case of Iran, among conservative political circles. In Iran, the oil ministry and NIOC support foreign investment as a necessary means to gain access to capital, but they have offered investment conditions and terms that do not satisfy the foreign oil companies. Political resistance to foreign investment is less of an issue in Algeria, where various political groups have come to terms with foreign companies' participation.

Upgrading Facilities

In terms of domestic industry challenges, marketing and refining experts in Kuwait and Iran mentioned the need to update export facilities. Both countries want to adapt their export facilities to the requirements of very large and ultra-large vessels and to increase the attractiveness of their facilities by, for example, matching the services offered in Saudi terminals.

Iran, Kuwait and Algeria all have plans to upgrade their refineries. Iran faces an important refining challenge. Its nine refineries, with a total capacity of 1.47 million b/d, are operational but aging (most were built before the revolution). Its refinery output is not adapted to domestic energy needs: too much heavy fuel oil (around 30 percent) and bitumen are produced; and gasoline is only 16 percent of production, forcing this major crude oil exporting country to import it. Iran imports approximately 160,000 b/d of gasoline, which it purchases at international market prices and sells at a subsidized price internally. Its plan is to reduce the production of fuel oil and to increase the capacity to produce gasoline while upgrading the environmental standards of the products. The main challenge, according to a director of the refining company, is that neither government nor private companies want to invest in refining because it is not lucrative.

In Iran and elsewhere, the deregulation of the subsidized domestic energy market is an important first step in attracting private capital and lessening the capital burden that falls on the national oil company and the state. In most countries, subsidies on energy are being challenged as economically unsustainable and are progressively being removed.[4] This change and the broader deregulation of the national downstream markets could increase FDI, which most oil ministries are eager to attract, though it is apparent to many in the region that the downstream in general is not a draw for investors. However, operating under more commercial terms should give incentives for the downstream companies to reduce their costs, which have hitherto been disguised by the transfer of subsidies.

The current trend is also given momentum by external pressures, in particular the requirements for joining the World Trade Organization. Saudi Arabia, for instance, has removed domestic subsidies for gas and has agreed to move toward adjusting all its domestic energy prices to international market levels as part of its negotiations for entry (although further progress is described as cautious and slow). Kuwait, an early member of the General Agreement on Tariffs and Trade, is said to have fully eliminated the subsidy on petroleum prices in 1999, with assistance from the IMF.[5] However, World Bank data on prices in Saudi Arabia and Kuwait indicate that domestically set prices remain below market value (see tables 7-1 and 7-2 for a comparison). Algeria has discovered that one of the last major impasses blocking its accession to the WTO is American and EU criticism of its artificially low domestic gas prices. And even though the United States has several motives for using

4. The contribution by John Mitchell at the end of this book makes the macroeconomic argument for phasing out subsidies in hydrocarbon-dependent economies.

5. "IMF Concludes Article IV Consultation with Kuwait," 4 April 2000, *http://www.imf.org/external/np/sec/pn/2000/pn0027.htm/*.

Table 7-1. Countries Pricing Super Gasoline below 37 Cents/Liter, November 2004

Country	*Price (cents/liter)*
Turkmenistan	2
Iraq	3
Venezuela	4
Iran	9
Libya	9
Myanmar	12
Yemen	19
Qatar	21
Kuwait	24
Saudi Arabia	24
Bahrain	27
Egypt	28
UAE	28
Oman	31
Algeria	32
Brunei	32
Trinidad & Tobago	35
Uzbekistan	35

Source: World Bank, 2005b: Table 5A.

its veto against Iran's accession, its objection to Iranian price subsidies is doubtless an additional hurdle.

Where there are subsidies or fixed internal prices for petroleum products, the financial burden rests on the NOC, which means that it has less money available for capital investments as well as that the transfer of oil rent to the state is reduced. In these cases, subsidy costs are included in the NOC's operational costs, which is frustrating for a company that would like to work along commercial lines. For example, NIOC is currently allowed to invest the

TABLE 7-2. Middle East Oil Producers Pricing Diesel below 37 Cents/Liter, November 2004

Country	*Price (cents/liter)*
Iraq	1
Iran	2
Libya	8
Yemen	9
Saudi Arabia	10
Egypt	10
Algeria	15
Qatar	16
Bahrain	19
Kuwait	24
Oman	26
UAE	28

Source: World Bank, 2005b: Table 5B.

money earned on exports of condensates and refined products and its share of buyback agreements (about $3 billion per annum), but part of this has to cover the rising cost of petroleum imports, which is currently about $1 billion per year. In many cases, NOCs lobby governments to obtain relief from the financial burden of subsidies. Governments may oppose these efforts, but in the end they pay for the subsidies through the NOC's expenditure. (If the Iranian government had to pay the fuel subsidy bill directly every month, it might be more inclined to convince parliament (or other relevant stakeholders) that the costs constrain other crucial public expenditure.) NIOC hopes that the reform bill being discussed in parliament will allow a complete overhaul of the subsidy system. A refining operations manager explained that "If we didn't have subsidies, we could develop better-quality products and we would have better operational costs. The subsidiaries' products are smuggled out of the country because they are so below the market that some people buy them here and sell them at international prices abroad." In the reform bill, Clause IV sets the price of energy for the domestic sector at market prices and the payments to be made between NIOC's subsidiaries for feedstock. The government would continue to support subsidies of natural gas, because of energy substitution targets that aim to free crude for export and generate more income in foreign currency. Subsidies for kerosene and liquefied petroleum gas would also continue because of concern for the energy-poor in rural areas. Each successive government would decide the amount of these subsidies and would allow for it in the budget.[6]

Providing cheap energy to Algerians puts Sonatrach under strain too. Ninety-five percent of its returns come from exports and 5 percent come from domestic sales. But volumes sold are not in the same proportions: 81 percent is exported and 19 percent is sold domestically (Sonatrach, 2004: 64). The subsidies take the form of fixed prices for the crude feed to NAFTEC (the refining subsidiary), for a number of products and for the natural gas feed to Sonelgaz (to produce electricity). The new hydrocarbon law will bring the domestic price of crude oil and its products and of gas to market value within four years and ten years respectively. Until now, Sonatrach has been the regulator of energy prices for refining, but under the new reform the state will take on this role.

6. The bill stipulates, for example, that subsidies will be allocated to provide a safety net for the poorest 10–20 percent of the population, to reinforce building codes (in view of the possibility of earthquakes), to improve public transport and to promote power generation from gas.

Supplying the local market takes priority over exporting oil for all NOCs in the study; and for all the governments concerned, providing affordable energy for the poor is clearly a political obligation that, as the central authorities of major oil and gas exporting countries, they cannot ignore. However, it appeared from comments made in interviews that many in the oil industry regard present subsidies to gasoline as really a subsidy for energy to the rich, those who own cars. Sonatrach is looking at experiments elsewhere so as to learn lessons about policies that can reduce subsidies while providing affordable energy to the poor. It sees Indonesia and Malaysia as examples to avoid and to follow respectively. In the former, violent protests over a sudden fuel price hike in 2003 almost brought down President Megawati Sukarnoputri's government and forced a policy reversal. The Malaysian government, by contrast, has taken a more gradual approach toward reducing rather than eliminating fuel subsidies, with much less political impact.

External Industry Challenges

Generally, external industry challenges do not appear high on the radar of most national oil companies. These challenges preoccupied most those working in marketing and refining; they were much less significant to those in upstream activities or even corporate planning. Sonatrach is an exception: its management is more aware of or concerned about external challenges. It is of course exporting more natural gas, which requires more long-term relationships with customers and stable conditions for new investment. It also has a declining domestic oil reserves base, to which it is responding with an internationalization drive. Table 7-3 details how the five NOCs perceive key external challenges.

Security of Demand

Marketing the hydrocarbons produced is the main external challenge to all the five producers; but because each produces different products, the nature of this challenge varies. Algeria's and Iran's main challenges relate to gas marketing, which requires long-term relationships in order to justify the initial investment. Iran is new to the scene, and its gas competes with supplies from established exporters. Kuwait's challenge is to find a market for its one, less desirable grade of crude, which is relatively sour (high in sulfur) and heavy (with a low calorific value). Algeria's and Abu Dhabi's marketing task is made easier by their lighter crude and their partnerships with internationally integrated oil companies, which can lift their production at their ports and sell it

Table 7-3. External Challenges—Perceptions of the Five NOCs

External challenges	*ADNOC*	*KPC*	*NIOC*	*Saudi Aramco*	*Sonatrach*
Security of demand	Important long-term issue	Important for sour/ heavy crude	Important for gas	Very important to maintain long-term demand	Important for gas
Environmental regulation	Strong concern about environmental impact	Concern (refining)	Concern (refining)	Concern (upstream)	Concern about environmental impact
Competition	Some concern over losing LNG markets	Concern about supermarket filling stations and IOC mergers (refining)	Concern over competition from Russian and Qatari gas	Confident	Concern about IOC mergers (refining) and Russian gas to Europe
OPEC supply restrictions	Confident	Confident	Confident	Confident	Confident

Source: Author's interviews.

on the international market. For Saudi Arabia, the mammoth supplier of 13 percent of the world's oil in 2004, the challenge is to find a home for each of those 10 million or so barrels produced every day. This task is helped by its multiple grades of crude,[7] which give it the crucial flexibility to sell on a variety of markets.

Gas exporters were more aware of markets. A Sonatrach corporate planner explained that the challenge is that gas as a fuel has not found a niche market in which, like oil in transport, it is irreplaceable. Therefore it is always in competition with other fuels. This explains why gas prices are tied to those of substitute fuels. The main threat to Algerian gas comes from the European Gas Directive, which has sought to increase competition between gas suppliers to Europe and to break up regional monopolies. Long-term bilateral risk-sharing gas contracts were the norm until 1998, after which the European Commission changed the rules of the game. Previously, most gas imports to European countries were bound by contract with the supplier to remain at the final destination of the product, which prevented their resale to another market in Europe. The EU has changed this destination clause, arguing that it has a common market and therefore that any energy exports to one European state can be transferred to another. This change is a major challenge to Sonatrach, which has seen its profit margins decline. It is forcing the company to respond with a strategy of increased integration and a more aggressive pursuit of markets. As a result, other producers are now clear competitors, and the company feels that it does not know its customers. Its managers are also resentful about being excluded from the decisionmaking process affecting their main market's fundamental characteristics: "The producing countries are unhappy about this. We want to be talking about complementarities rather than security of supply [which is what the EC worries about]."

Iran's gas development challenge pertains to its late entry into the gas market. Qatar has already established itself as an important supplier of LNG to Asia. Russia and Algeria are key suppliers of pipeline gas to Europe, and Sonatrach hopes to expand its LNG business in the United States. This well-established competition is clearly a concern for Iranian oil officials; but on the other hand, the country's strategic position gives it an advantage over Qatar in that it has the potential to develop a pipeline network to South Asia and, ultimately, toward the Mediterranean too.

7. Saudi Arabia has five grades: Arabian Heavy, Arabian Medium, Arabian Light, Arabian Extra Light and Arabian Super Light.

As stated above, Kuwait's specific marketing challenge is that it has to "find a home for its heavy-grade crude," which faces declining demand. As an international marketing director at the Kuwaiti company explained, it used to be easy to market Kuwait's crude and its products. Now, consumers are more aware of their need for security of supply and for more environmentally sound products. As a result, they build their own refineries. India, for example, was an important importer of diesel until the late 1990s; now it exports this product. It is therefore more difficult to find new markets. Though the Kuwaiti oil industry has long been integrated, it is undergoing a transformation of its business activities involving the use of increasingly sophisticated marketing tools.

Saudi Aramco's specific marketing challenge—that it has so much oil—will not disappear. Possessing an estimated 23 percent of the world's oil reserves and currently producing 13 percent of total supply, it has room to expand production but would need to find markets for that additional oil. It has to ensure that demand for oil continues for as long as Saudi Arabia has oil underground. Maintaining oil as a primary source of energy over the long-term is a key challenge that Saudi Aramco takes very seriously. It is investing more and more in the research and development of new technologies that will sustain demand for oil. A scientist at Saudi Aramco explained the company's efforts to use science in support of future oil demand: "We want to protect our oil market, through fossil fuel-driven fuel cell technology for instance. We would like to do this with NOCs, with OPEC producers. We have common concerns."

The president and CEO of Saudi Aramco has indeed talked about the company's interest in developing NOC–NOC collaboration in order to develop clean-oil technologies.[8] Saudi Aramco approached KPC about establishing a joint investment program for fuel cell research as part of its effort to sustain long-term demand for oil. One executive involved was disappointed by KPC's response: he claimed that the Kuwaiti company has a more commercial approach to the problem (seeing that and other investments as for profit only) and lacks the strategic vision necessary for influencing the course of future research. Saudi Aramco has invested in joint research on fuel cell technology with Norway's Statoil instead.

Saudi Aramco's top executive has also spoken of the company's interest in conducting work jointly with other producers on regulatory and taxation issues that affect oil demand. For most oil producers, these issues center on

8. Presentation by Abdallah Jum'ah (see note 3 above).

the huge amount of taxes that governments of consuming countries impose on oil. In 2002, EU taxes on oil consumption were 59 percent of the landed cost of the "composite barrel" of imported crude (OPEC, 2003: 120). In fact, the taxes fall almost entirely on transport fuels (on which the tax in Europe is two to three times the cost of the product) rather than on fuel oil for industry and power. In 1999–2003, the G-7 countries collected $1.4 trillion in taxes from oil consumers, compared to OPEC's revenues (including costs) of $1.1 trillion (OPEC, 2004).

Governments of importing countries often forget how annoyed the producers are that they are making more money from taxing the product than the producers are for producing it. A French oil executive told me a story that illustrates this: French Minister of Exterior Commerce François Loos went to Qatar and met with Minister of Energy and Industry Abdullah bin Hamad Al-Attiyah. The French minister apparently started to complain about high oil prices. Al-Attiyah asked, "Would you like us to ship oil free to your ports in Marseilles?" Loos, flustered, stumbled, "Well, yes!" "OK, just split the taxes with us 50-50 and we'll do it." Al-Attiyah is a clever man, but taxes are an irritant for producers that is unlikely to be removed. Iranian oil officials have argued that some of the revenues from taxes on petroleum should be channeled back to investment in new capacity in the low-income oil exporting countries. It is unlikely, however, that any of these suggestions will be taken up.

Environmental Regulation

Commitments in consumer countries such as to reduce carbon emissions, set environmental standards for hydrocarbons and promote alternative fuels naturally have an impact on exporting countries. However, the level of concern about these external challenges among the five NOCs under study is unequal.

Saudi Aramco contributes scientific research in response to these challenges. In order to reduce its own emissions and to improve the image of oil, its research center has a program on carbon sequestration (capturing emitted carbon and storing it) and is investigating means of capturing gas flared at oil fields for reinjection into oil reservoirs. It is conducting research as well on innovative ways to desulfurize petroleum. These steps are clearly important, though it seems that many producers lack an understanding of how they can benefit from proactive action on this and other environmental fronts, notably on the climate change regime, emissions trading and energy efficiency programs.

In 2004, all the companies under study showed a lack of interest in the impact of the Kyoto Protocol and climate negotiations on future demand for oil and gas, even though managers often discussed the importance of thinking strategically about long-term threats to oil demand. There was little awareness of this issue, except in Iran, which had already made commitments to the protocol as part of the Asian group. One long-term planning manager at Sonatrach felt that as long as the United States and Russia do not ratify the agreement, "it does not exist." Since those interviews took place, Russia has ratified the Kyoto Protocol—as did the UAE, Saudi Arabia, Algeria and Kuwait in early 2005. Awareness of this strategic issue and responses to it are therefore likely to surface.

The challenge the Kyoto Protocol poses to the oil industry is likely to become more substantial over the next few years. Energy producers will have to consider how they will be affected by the new climate regime that evolves from the Kyoto negotiations and how they could benefit from it. For their part, as Jacqueline Karas has noted, climate change experts have not appreciated the role of producers in that regime and the likelihood that greater participation by producers in the negotiations could be mutually beneficial.[9] For example, companies that might otherwise be deterred from the heavy infrastructure spending involved in developing means to capture flared gas for domestic use or other emission reduction measures could join forces with other producers (private and national) to share research and development costs. As Statoil has argued (at the 2005 National Oil Company Forum in Brazil), the companies' spending share could be converted into emission credits approved under the Kyoto Protocol mechanisms.

The impact of consumer country legislation on limiting emissions is felt more immediately in downstream activities. Refiners and marketing experts at KPC, KPI and Iran's NIORDC, the refining company, talked more precisely about today's challenges of environmental regulations; they mentioned European environmental regulations on gasoline, which put pressure on their costs. In terms of quality of refining capacity, both Kuwait's and Iran's products are not up to European specifications. Iran planned to adopt these specifications by 2005.

ADNOC was unusual in its concern for the environmental impact of the oil industry. Abu Dhabi's leadership has sent a strong signal on this matter, raising the importance of energy conservation and environmental protection

9. Comments made at the "International Energy Dialogue," a Chatham House–International Energy Forum workshop, London, 27–28 April 2005.

to a high strategic level. Almost every person interviewed mentioned the company's relevant achievements and the continued importance of the issue. However, like its counterparts in the Middle East, ADNOC headquarters did not appear to switch off its lights after work hours. By contrast, the Petronas Twin Towers in Kuala Lumpur stood in almost complete darkness at night, and many light switches were on timers in the building during the day.

Competition

Those interviewed generally did not like to talk about competition, and most tended to minimize the threat to their market share from other producers. All the same, oil must compete against oil and against other fuels in the global energy market. NOCs are reaching out closer to the end-user, which puts them in competition with international oil companies. IOCs have an advantage in linking the upstream and the downstream internationally; national producers do not have the same financial, technological and commercial advantages in gaining access to markets outside their borders (Chevalier, 1994: 13). Challenges also emerge from neighboring producers, most notably Iraq. Political and industry developments in Iraq will affect market shares, the policy context and the investment climate for neighboring states.

In terms of downstream competition, KPC's international refining subsidiary KPI discussed its concern about competition from the growth of self-serve filling stations and hypermarkets (large supermarket chains) in Europe that can perform with low margins. It also perceived a challenge in the trend of industry consolidation, which means that its competitors are leaner. "Fusions, alliances and new monopolies" in the oil industry were also a concern at Sonatrach. In response to the assumption made by its industry peers that KPC should be making huge profits in the downstream because of its access to large reserves of cheap oil, one of its marketing directors explained that the Kuwaiti government reaped the profits from the low costs of production and that this was not an advantage for the national oil company: it buys the crude oil from the state at market prices before refining it and selling the products. It must therefore work carefully to sustain its competitiveness.

KPI's ability to meet this challenge was questioned in the company, and KPI acknowledged that it has had difficulty in developing new international refining capacity. In order to improve Kuwait's access to international refining markets in growth areas, KPC and KPI had sought to acquire refineries in India and China. Their bids were unsuccessful for various reasons, including the difficulty of dealing with informal channels in the host countries, which required a good intermediary—an irony of sorts, because investors face the same difficulty in the Gulf. Also, some acquisitions were made before the Asian crisis of 1997 in expectation of high returns, and had to be sold. A KPI executive regretted that the company had not found an IOC with local market access with which to form a partnership, as had Saudi Aramco. But in spring 2005, KPC announced that it would seek new ventures in Asia with BP and in China with Shell.

On competition for crude oil markets, there was much ado when most interviews took place (2004) about the imminent rise of Iraqi and Russian petroleum exports. They would threaten the market share of producers in the Middle East and there would be a scarcity in the region of technology, equipment and capital, which would be expended in developing other petroleum resources. There was also much talk about the threat of non-OPEC oil internationally; the volume of its production was forecast to rise over the next 5 to 10 years until it reached a peak, reducing OPEC's global share of the market in the meantime. The substance of these forecasts has changed dramatically since then. Iraq's production increases have been very modest, primarily as a result of attacks by insurgents against facilities and pipelines but also because of the very slow emergence of a legitimate, sovereign regime and the general level of insecurity in the country, which have put off foreign investors. The developments on Russia's energy front, specifically the dismantlement of Yukos, the most successful Russian oil company, for arrears in taxes owed to the state, have thrown cold water on the plans of most IOCs to follow in BP's footsteps and invest in Russia. But before these developments, observers of the energy scene in early 2004 (including myself) saw a challenging short to medium term for Middle Eastern producers, full of competition. This would be followed in the long term by a comfortable rise in market share. Also, crude oil prices were already high, with an average spot market quotation of $30/b for Brent in January 2004, but nowhere near the $50–70/b that followed later in 2004 and 2005. Most of the comments made regarding international competition from other producers must be seen in the cautious context of early 2004.

At Saudi Aramco, I was surprised to find a lack of concern about the projected rise of non-OPEC oil production over the next 5 to 10 years and the possible loss of Saudi market share. The feeling in the company appeared to be rather philosophical: "We've weathered these things before; we will do so again." Saudi Aramco is in the game for the long term; and though the oil ministry is concerned with short-term threats to the oil price, the company disregards the medium-term threats and focuses on the long-term evolution of its environment. Respondents at Saudi Aramco tended to see external challenges in terms of long-term threats to oil demand, as discussed above.

As for specific competitors, it also appears that Saudi Aramco, like other Middle Eastern producers, did not perceive Iraq as a threat to its market share or to the balance of OPEC. A number of industry commentators and journalists had at the time depicted future Iraqi oil production as a threat to the large exporters in OPEC. But most professionals in the region felt that there would be room for all and that Iraq would stay in OPEC and that its oil policy would not change dramatically. They also responded positively to requests for support made in 2004 by interim Iraqi oil minister Ibrahim Bahr Al-Ulum. In Kuwait and Saudi Arabia, he sought help in particular with drawing up plans for the future structure of the Iraqi National Oil Company and the national industry's operations. He solicited studies from KPC's and Saudi Aramco's corporate planning departments and from colleagues he knew. Surprisingly, the most concrete support came from the Kuwaiti oil industry and oil ministry. Of all the Gulf governments, the Kuwaitis have collaborated most closely with the Iraqis since the American occupation of Iraq. In January 2004, the governments of Kuwait and Iraq formed a joint committee that decided Iraq would provide natural gas exports to Kuwait. In turn, Kuwait would allow the use of its export facilities for Iraqi oil; Kuwaiti companies would help with the services and material needed for the day-to-day maintenance of Iraq's oil fields and develop existing fields in Iraq in a joint venture with INOC.

When I expressed surprise to Kuwaitis at the rapid thawing, and even warming, of relations with Iraq, they explained that their societies were already intertwined. A Kuwaiti executive explained that one third of Kuwaitis have family in Iraq and a number of them own agricultural land there. Also, Kuwait owes its current economic boom in good measure to the war and reconstruction in Iraq. "Business interests run thicker than blood. The prospects of making money make everyone forget the past," she explained. It is a good basis for more cooperative political relations.

Strangely, short memories appeared to dominate in this region with a long history. Following Kuwait's official apology to Iran for supporting Saddam Hussein in the Iran-Iraq war of 1980–88, it appears that the Iranians have forgiven their neighbor. A senior figure at the National Iranian Oil Company told me that after Kuwait's liberation, a high-level delegation from Iran visited the country to offer support. The emir told his guests that Kuwait was grateful for their generosity in breaking the "bad relations." Also, Iran has been seeking to improve relations with the Iraqis since the removal of Saddam Hussein. A number of people in Iran's oil industry spoke generously about the Iraqis, whom they felt were suffering. From these conversations, I gathered that there is little rancor left from events of the past 25 years, except between Iranians and Saudis. In Tehran, it was drawn to my attention that the Saudis had never apologized for supporting Iraq in the war of 1980–88 or for the killing of 402 Iranian pilgrims in clashes with Saudi security forces at an Iranian-led rally in 1987—a crisis that led Iran to boycott the Hajj in 1988–91. One senses that Saudis remain dismissive of Iranians, whom many Wahhabi Sunnis consider to be infidels because of their Shiite faith. Simplifying the Iranian view, Saudis are socially backward religious extremists, and their own society is open and modern.

There was good will between Saudi and Iraqi officials, but the Saudi government was slow to respond to the Iraqi requests for advice (see above). An oil ministry official said in February 2004 that "We are engaged now. We had discussions internally about whether we should [respond] while Iraq was still under occupation. We decided we must do so now because we don't know when the USA will be gone." The Saudi oil industry is assisting its Iraqi counterpart with technical cooperation, reservoir management, gas injection, water intrusion and reservoir pressure. But because of the serious security situation in Iraq in 2004–05, most of this assistance was given from outside Iraq.

Generally, I perceived a great deal of respect for Iraqi oil professionals. Many of their counterparts in the Middle East felt that Iraqis are "the most talented in the region." "We all learnt from the Iraqis," explained an influential Saudi official. Such admiration is hardly usual in the region, where most are dismissive of the competence of their neighbors.

In 2003–04, reports abounded in the press predicting a clash over crude oil markets between Russia and the large Middle Eastern exporters, especially Saudi Arabia. A common view of commentators and analysts (before the Yukos affair and the increased perception of risk for foreign investors in Russia) suggested that Russia's vast potential would challenge the Gulf's market share, reduce OPEC's capacity to influence prices upward and weaken Saudi Arabia's strategic importance internationally. Moreover, with Russia contemplating whether, how, when and where to export its oil and gas to northeast Asia, a common perception was that its hydrocarbons would challenge the Middle Eastern suppliers in Asia, their traditional and most lucrative market.

In fact, interviews revealed that many viewed the threat of Russian production as overstated and "drummed up." They were certain that Russian reserves had been damaged when the Soviets pushed their oil fields too hard in the 1980s. An Iranian planning ministry official added, "They are producing at capacity. Anyone who is not producing at capacity now with these prices is an idiot! . . . If they weren't at capacity, the Russians would be a threat." A Saudi Aramco corporate planner felt that "the bigger threat is if Iraq is pressured to develop very fast, bringing . . . a lot of oil on the market in a short time. There's a threat because the reserves *are* there." But the vast majority interviewed in the Middle East thought that Russian oil and gas would not challenge their markets because demand would be high. A senior international affairs specialist in Iran added that with the Asian markets growing, there would be room for all. The following two comments from the Gulf echo this view: "We don't see them as taking our markets. If they don't take them, we will have to do it! And we don't want to rush our production. We treat our reservoirs like a person, like a baby [and handle them with care]." "The threat is there, but it's not serious. Russia wants to export to Europe, to the USA, but in Asia, there will be a delay. It won't happen for a long time. Now there is only trucking [truck convoys to China]."

The general feeling in the region appeared to be that Russian oil and gas constitute only a short-term threat. As large reserve holders, the Middle Eastern producers are around for the long term and will outlast Russian production. Iranians, like Saudis, are dismissive of Russian reserves relative to theirs. In terms of gas exports, the competition from Russia is a greater concern, as a comment from an Iranian planner demonstrates: "They are a threat to gas markets, though. We should divide the market with the Russians. We supply Asia and they supply Europe. If they don't interfere in Asia, we don't interfere in Europe."

As for Algeria, it has historically enjoyed close ties with Russia. Algeria's socialist system was inspired by the Russian model, and the two countries have developed a great deal of technical cooperation. Russian companies have also been involved with Sonatrach in building pipelines in Algeria. In the past, the two countries have done market swaps for natural gas.

However, the commercial relationship between Russia and Algeria has changed owing to the European Gas Directive: Russian gas is now perceived as the main competitor to Algerian gas. The European gas market is central to both countries. Algeria sends 95 percent of its exports to Europe.[10] In 2000, natural gas exports provided 16 percent ($16.6 billion) of Russian and 33 percent ($7.1 billion) of Algerian convertible currency revenues (Stern, 2002: 19). But Russia and Algeria can no longer control the destination of

10. *http://www.mem-algeria.org/actu/comn/doc/competition_florence.htm.*

their gas within Europe, because it can be resold from one country to another. As a result, Russian gas could reach markets in southern European countries and undercut Algerian gas. However, there is at least some potential for this EU-driven competition to push Algeria and Russia together in an effort to counter liberalization via some kind of informal OPEC for gas. In March 2002, Russian and Algerian energy ministers set up a joint working group to conduct a dialogue with the European Union on gas exports; the Russian minister spoke of coordinating the export policies of the two countries. In 2001, they participated in the creation of the Gas Exporting Countries Forum, made up of 11 gas-producing states, including Iran, Qatar, the UAE and Norway (as an observer).

Qatar is also a dominant gas exporter, and its market dominance worried one Iranian planner: "We can't compete with Qatar. We look for markets where Qatar is not able to get easy access, India and Pakistan, for example, where we have land access and the Qataris would need deep water pipes in the Indian Ocean and the Oman Sea. . . . We can't compete with Qatar's prices. They sell LNG at almost the same price as our pipe gas." But other planners in Iran felt that demand would exceed Qatar's supply. They explained that they had taken into account Qatari competition, for instance when studying the feasibility of gas development plans in South Pars.

ADNOC too saw Qatari gas, with its "one field, one pipeline and one type of gas," as a tough act to beat. Even so, it does not appear to see Qatar as a threat because it has to some extent surrendered (for now) and will import gas through the Dolphin project—an integrated natural gas pipeline grid for Qatar, the UAE and Oman. Its own gas export ambitions are limited as a result of the strong growth in domestic demand.

In the gas sector, there are a relatively limited number of players, and each exporter's market is more clearly defined. A Sonatrach planner presented the situation as follows: first, there are the traditional gas players, which include Qatar, Indonesia, Malaysia, Brunei and Algeria. Then come the new players, including Australia, Oman and Nigeria; and finally, there are the future players, Venezuela, Brazil, Angola and Mexico. The traditional exporters have already divided up the markets, and the latter two groups are more clearly Sonatrach's competitors because they are trying to penetrate markets that it has already marked out for itself (the American market) or currently occupies (Europe). As pointed out above, 95 percent of Algeria's gas is exported to Europe; the remaining five percent goes to the United States and Tunisia. With European countries' attempts to diversify imports, Algeria must seek out new markets, and thus will increasingly compete with other producers.

Table 7-4. The Five Producers' Oil-Production Capacity, Production and OPEC Quota, Third Quarter 2005 (million b/d)

Country	*Production capacity*	*Production average*	*OPEC quota*
Algeria	1.380	1.366	0.894
Iran	4.000	3.944	4.110
Kuwait	2.600	2.529	2.247
Saudi Arabia	10.500–11.000	9.501	9.099
United Arab Emirates	2.500	2.474	2.444[a]

Sources: OPEC Monthly Oil Market Report, November 2005; EIA Short-Term Outlook—November 2005, Table 3a.

Notes: Crude oil does not include lease condensate or natural gas liquids. OPEC quotas are based on crude oil production only. Capacity refers to the maximum sustainable production capacity, defined as the maximum that could be brought online within 30 days and sustained for at least 90 days. Kuwaiti and Saudi figures each include half of the production from the Neutral Zone between the two countries. Saudi production also includes oil produced from the Abu Safa field on behalf of Bahrain.

a. This quota applies to the emirate of Abu Dhabi only.

OPEC Supply Restrictions

I asked the five producers how they managed new exploration, new reserves and new production capacity within the constraints of OPEC quotas. Surprisingly, none of them expressed concern over the capacity of OPEC to allow for quota adjustments and, specifically, over its acceptance of an increase of their own quota. In Algeria and Saudi Arabia in particular, actual production levels are mismatched with existing OPEC quotas. In the third quarter of 2005, OPEC countries were producing at close to maximum capacity—except in Saudi Arabia (see table 7-4). But under a less strained market environment, the pressure would be reversed, as producers would be required by OPEC decisions to cut back their production in line with their quota. Should the producers develop their production capacity beyond their quota, they would be left with idle capacity in a normally supplied market. Both Iran and Algeria have pressed OPEC for quota increases. Algerian oil officials and professionals appeared confident that their bid to increase their quota would be approved. Iran's oil ministry officials felt that OPEC quota restrictions were only a short-term problem and that those producers with idle capacity would hold all the cards.

Another crucial issue for the members of OPEC is how they will deal with the pressure of some producers' need to increase their hydrocarbon export revenues so as to meet the growing needs of their societies. A senior figure at KPC felt that the short-term budgetary needs of states such as Saudi Arabia, Iran and also Iraq were indeed a core issue. Those needs would lead them to limit production and increase revenues by way of high prices. At the Iranian Ministry of Planning, the analysis was different: "For Saudi Arabia and Iraq, there is no choice but to pump as much as they can. For Kuwait, Iran and the UAE, there are other choices." But whether they agreed with the Kuwaiti manager that Saudi Arabia's strategy would be to hold back on production and keep prices high or with the Iranian official, who thought that it would push production in order to keep revenues coming in, all agreed that Saudi Arabia is the only producer that can play with both price and volume.

Outside Saudi Arabia, there is a feeling that the producers' ability to control prices is declining. Within marketing departments, most respondents felt that as producers they were limited in their means to influence the crude oil market—except when supply is tight, when incremental supply can make a difference. Kuwait and Iran, for example, do not have spare capacity or sufficient volumes to play with the price of oil. But in Saudi Arabia, the tone is quite cautious about its power to influence markets, as this extract from a conversation with a senior vice president at Saudi Aramco illustrates:

> VM: "If others develop spare capacity, Saudi Arabia is the first to pay because it won't be able to use its own spare capacity."
>
> SA: "We don't build spare capacity [solely] to make money on it. We are citizens of the world. We do this to stabilize the markets. People are not seeing this. We are meeting our obligations."

Conclusion

With regard to strategic market challenges such as competition from other producers or from other fuels, there was a generally low level of interest among the five NOCs discussed in this book. Their confidence may be well founded, but I wondered whether it was to some extent a result of the NOCs being insulated within their borders, distant from global industry concerns. These national oil companies operate mainly in the upstream of their countries, in contrast to IOCs, which have interests across the globe that help them to stay close to markets and opportunities. For instance, KNPC, the Kuwaiti refining subsidiary, buys crude from KPC and sells the refined products back to it. As a result, it does not have direct contact with customers.

The NOCs' international contacts are restricted in most cases to their marketing departments, international subsidiaries, operating companies (in the cases where they are in partnership with foreign companies) and international representative offices. These contacts are their antennae, a crucial link between the companies and the rest of the world. In this respect, there is a worrying trend: some of the NOCs studied here have closed their international offices—as demonstrated by the closure of international offices of KPI and of NIOC's Kala Ltd. (a procurement company). The closure of Iranian international offices follows a decree requiring state companies to comply with cost-cutting measures. ADNOC does not have offices abroad.

Some observers in the NOCs felt that international offices do not add value. Their assumption was that new communication technologies would allow them to stay close to customers and markets from the comfort of their headquarters. However, an expanded vision of these offices' mission would see them as important antennae for current trends and different views. It appears that Saudi Aramco is taking a more strategic view of them—this is reportedly tied to the rise of new executives within the company who want to see the company exchange views more often and more widely with outsiders. Its London office, Saudi Petroleum Overseas Ltd. (SPOL), has expanded the scope of its activities significantly, in order to study supply and demand trends and the political forces that affect them.

eight
The Home-Front Strategy

Strategic thinking is a new imperative for the national oil companies of the Middle East and Algeria. It is the result of pressure on available resources, the need to do more with less, and their new drive to succeed as commercial organizations. The NOCs' strategy focuses on optimizing oil and also gas resource development, pushing the integration of their activities further, supporting the national economy and moving toward greater internationalization. This chapter will examine the domestic strategic focus of the five NOCs under study and their ability to deliver on the above objectives. Their plans for internationalization will be addressed in the next chapter.

Reactive or Proactive Strategies?

Timing is everything in the oil business. The global oil market is dominated by a continuous process whereby demand surges, which strains supply and brings high prices. This leads producers to invest in new supply, which, with some delay, enables new oil to flow that deflates prices and stimulates demand once more. The process will have come full circle. Well-timed investments pre-empt surges of demand and make new capacity available when the market is tight and other exporters are producing at their maximum capacity. Additional supply can be sold at higher prices.

Well-timed investments require good forecasts of future supply and demand. Most of these forecasts are prepared by government agencies in the oil importing countries, for example the International Energy Agency (IEA) and the American Energy Information Agency (EIA). These countries are concerned, above all, with security of supply: they want more oil to be available at lower prices. The oil producers, national and private alike, concentrate on security of demand: they want to make sure that consumers will pay them well for the oil they produce. The importing countries' forecasts have a poor track record in predicting surges and slumps in demand. They focus on either the short term, up to 18 months, or the long term, 20 years away, thus avoiding the questions that are critical for the timing of new investments. A similar criticism can be made of the OPEC Secretariat's forecasts. Private consultancies also prepare forecasts, but these have not been reliable either. The oil majors produce their own forecasts, which they do not make available to others. They have teams of economists who base scenarios of future trends on various economic and political assumptions about available supply and demand. Some of their executives have conceded in interviews that these forecasts failed to notify them early on of China's economic growth, which has contributed to a prolonged period of constrained supply.

As for national oil companies, they tend to rely on data prepared by the IEA, the EIA and the OPEC secretariat. Some can supplement these data with analysis by their corporate planning department, but others, such as ADNOC, do not have a team of economists. This inhibits their capacity to develop long-term strategy and makes their investment program reactive to market forces. In fact, I was surprised to find that strategic planning was new to the Kuwait Petroleum Corporation and the National Iranian Oil Company. A Kuwaiti executive explained that "Before liberation, KPC had no strategy. People were just relaxed. Not really worrying about things. Not really thinking. KOC pioneered strategy for the company. It took two years to get their strategy and they just presented it to KPC. They were surprised. It led to KPC making its own strategy. But it isn't so defined, or so clear. We need an action plan."

The strategy-making process at Sonatrach was revised recently in order to respond more effectively to industry challenges. It now makes projections over 20 years and plans for five years. Saudi Aramco has strong, well-established planning processes. It prepares a business plan every year covering strategic investments over a rolling five-year horizon. Saudi Aramco and ADNOC also prepare contingency investment plans, to be better prepared when requests come from the Supreme Petroleum Council. Regarding plans for expanding its production capacity, for instance, ADNOC has not yet received directives from the SPC, but it is preparing itself: "Part of our long-term strategic plan is to be ready to produce more if the government wants more. I am ready. I send out for seismic surveys to specific fields now [to be prepared]. If the data coverage is insufficient, we will make the expense [of ordering the seismic surveys]. This gives us a one- to two-year head start. But we won't invest in raising the capacity until they say. We have a priority list of different fields, with duration and costs [that we have analyzed already]." An E&P director explained that when the signal from above is sent to raise production capacity, the new fields can be brought into production in five to six years. The company has plans ready to present to the government (through the SPC); they give different options involving more or less certainty and cost.

Planning is an area about which a number of leading IOCs offer advice to NOCs. As a Western oil executive in the Emirates explained, "[We are] trying to instill the need for planning. Careful planning is always a central concern for IOCs, who do not want to invest in production expansion without careful consideration of future supply and demand projections lest they should find themselves with excess capacity."

For very large national exporters, "excess" capacity is actually a desirable outcome, and is more positively termed "spare" capacity. It gives the producer a time buffer in which to respond to increases in demand, and the negative effects during periods of weak demand can be limited by OPEC quotas. For instance, when there was an unexpected, and indeed unprecedented, increase in global oil demand in 2004, Saudi Arabia made most of its existing spare capacity available so as to keep the market supplied. Meanwhile, Saudi Aramco duly upgraded its production capacity to 11 million b/d, restoring at the same time the kingdom's strategic minimum reserve capacity of 1.5 million b/d. Saudi Aramco expanded capacity by 800,000 b/d by developing the Qatif and Abu Safah fields and drilling new wells elsewhere. In this case, 500,000 b/d of new capacity were made available in only six months because contingency plans had been prepared for a surge in demand. Though maintaining spare capacity makes the producers less dependent on forecasts, it still requires planning.

From interviews, it appeared that Middle Eastern oil producers were worried about developing too much capacity too soon. Before the recent high prices, there had been two decades of surplus capacity. Asked in January 2004 about the expected growth in demand for OPEC oil over the next two decades, marketing specialists in KPC felt that it would be marginal for the next five to ten years. A KPC director of international marketing concluded that "There are bad times coming for OPEC." By the end of 2005, despite two years of sustained high prices and the evident resilience of Chinese demand

Table 8-1. Upstream Investment Targets of the Five NOCs, 2004–05

NOC	*Production (3rd quarter 2005 average)*	*Production capacity (million b/d)*	*Production capacity target (million b/d)*	*Time frame*
Algeria	1.366	1.380	2.0	by 2010
Iran	3.944	4.000	5.4	by 2008
			7–8	by about 2025
Kuwait	2.529	2.600	3.5	by 2008
			4.0	by 2020
Saudi Arabia	9.501	10.500–11.000	12.5	by 2009
United Arab Emirates	2.474	2.500	3.0	by 2006
			3.7	by 2010

Sources: 3Q05 b/d averages calculated from monthly averages given in OPEC Monthly Oil Market Reports, October and November, table 16 (OPEC figures are based on secondary sources). For notes on capacity and production, please see notes to table 7-4 in chapter 7. Investment targets from author's interviews, NOC literature, official government sources and EIA country reports, 2005.

for imported oil, interviews with the NOCs reveal their concern that they have been enjoying a bubble likely to burst at any time.

This worry, that the cyclical price pattern will shortly bring a downturn in demand, has not prevented all five producers from going ahead with ambitious plans for increasing their production capacity in the short to medium term, as shown by the targets in table 8-1.

Expansion Plans

All the producers under study have plans to develop spare capacity. Saudi Arabia, as noted, intends to maintain 1.5 million b/d of spare capacity. It does so despite high costs because its policy is to keep the markets well supplied and prices stable. In this way, the kingdom sustains long-term demand for oil and maintains its external sources of support. In view of the projected strong demand for energy, and perhaps as a result of external pressure to raise output, oil minister Ali Naimi announced in 2005 that Saudi Arabia would double its spending plans for the energy sector to around $50 billion for the next five years. (The target is net of expenditure on maintaining facilities.) In addition to plans to add 800,000 b/d of refining capacity to its existing 3.3 million b/d, the investment target of $50 billion includes a number of production-boosting projects; these would offset the natural decline of oil flowing from the kingdom's present fields and increase pumping capacity. Saudi Aramco's objective is to maintain 1.5–2 million b/d spare capacity above higher production targets, leading to a maximum capacity of 12.5 million b/d by 2009, and eventually to 15 million b/d, while it replaces reserves produced.

Some of the replacement will come from improving recovery factors.[1] Saudi Aramco, like other Middle Eastern producers, emphasizes recovery factors in its development strategy. Dr. Nansen Saleri, Manager of Reservoir Management at Saudi Aramco, has presented figures that show that the company plans to extract more than the industry average of around 50 percent from reservoirs. It hopes that with evolving technologies, it can extract up to 70 percent (Nasser and Saleri, 2004). According to one of the company's reservoir engineers, depletion rates are at one to two percent. This highlights the importance of optimizing the development of existing reserves in the company's strategy. Maximizing the recovery of hydrocarbons from existing reservoirs while keeping the depletion rate low is perceived as an obligation to future generations. It means that producers do not push development in order to maximize the hydrocarbons extracted today but cradle their reservoirs "like a baby" so that they maximize extraction over the life span of the reservoir. No company takes this more seriously than Saudi Aramco, guardian of the world's largest reserves. Executives at the Kuwait Oil Company and ADNOC's operating companies also emphasize the importance of recovery over production. These producers are far from hitting their peak, and they want to maintain a cautious ratio of production to reserves.

Algeria is in a different position because it is approaching the limits of its national reserves base. Sonatrach plans to boost output, draw in increased investment and expand its own overseas operations. As for national oil policy, the Algerian government wants to see an aggressive exploration drive, carried out mainly by foreign companies with the participation of Sonatrach. Its policy is to continue to involve international capital and management through a variety of mechanisms that have so far proved to be attractive to major foreign companies. Regional banks estimate that, after Saudi Arabia and Qatar, Algeria will see the greatest acceleration in energy investment in the region in 2005–09, at around $20 billion (Aïssaoui, 2005). The Ministry of Energy and Mines has set a production target of 1.5 million b/d for 2005, which had not been reached at the time of writing—maximum sustained capacity was 1.38 million b/d, and production averaged 1.36 million b/d in the third quarter of 2005. The ministry is aiming for an output of 2 million b/d by 2010. However, Algeria's OPEC quota is currently set at 894,000 b/d, which means that the country either will have spare capacity or will produce above its quota until it renegotiates its quota allocation. For now, this ques-

1. Recovery factors are used as a target figure for how effectively a field is drained. Producers seek to improve the proportion of resources which can be recovered from a deposit to the resources originally in place.

tion is on the back burner because OPEC members decided in 2004–05 to ignore formal quotas and produce to capacity. However, should market tightness be reduced and prices fall, tension within OPEC will resume, as producers will be called on to rein back production more closely in line with quota allocations. Algeria will not want idle capacity: its volume of production is not large enough for this supplemental capacity to translate into international political influence or to be used effectively to influence prices. As mentioned in chapter 7, spare capacity is also unlikely to please foreign investors, but energy ministry officials are adamant that increasing quotas will not be an issue and that foreign investors agree about this.

Iran too has a production expansion strategy. In numerous interviews with high-level managers and officials, it appeared that Iran wants to take on a role in the oil and gas market (and thus in international politics) that is in line with its important reserves. Its objective in 2004 was to raise production capacity from 4 million b/d to 5.4 million b/d by 2008 and to 7–8 million b/d by 2025 and to replace the depleted resources. A senior official at the planning ministry put the total investment needed to achieve the 7–8 million b/d target at $104 billion, a sum that includes investments in refining and petrochemicals. However, the national oil and gas industry in Iran is capital-constrained because funds are needed to support the non-oil economy. The Iranian Ministry of Petroleum has responded to these challenges with a dual policy: first, lobbying parliament for the removal of subsidies on gasoline and, second, attracting foreign investment in order to increase its upstream oil and gas capacity.

Kuwait has set ambitious targets for expanding its production capacity: 3 million b/d by 2005, 3.5 million b/d by 2010 and 4 million b/d by 2020. But by the third quarter of 2005, Kuwait's output only reached 2.5 million b/d. Project Kuwait is expected to contribute substantially to further expansion of capacity. The project involves the participation of foreign oil companies and $7 billion of upstream investments in the northern oil fields. The project has seemed to be going nowhere as a result of political opposition from members of parliament who resist the opening up of the oil sector to foreign investment. However, as we saw in chapter 4, there is hope for a resolution of the standstill, as the Ministry of Energy has modified the project's terms in response to the concerns of its opponents.

Abu Dhabi's policy is to maintain spare capacity in upstream oil. Like Saudi Arabia, the UAE has acted as a swing producer within OPEC, compensating for supply shortfalls since 2002, and it can boost capacity quickly at its onshore fields. Despite its large reserves, its expansion plans are relatively modest. Abu Dhabi intends to build a sustainable output capacity of around 3 million b/d by 2006, up from 2.5 million b/d in 2005. ADNOC plans to spend $1.5 billion a year over the next four years on capital projects, of which at least 40 percent are in the oil sector, in order to boost capacity at existing fields and to develop new finds. It has devised plans for additional increases in production capacity above this target. The SPC has not yet directed the company to implement these plans, but it appeared from my interviews in the company that consultations were well under way. It also plans to produce a greater volume of gas; this will be used, as now, for reinjection into oilfields.

Capacity to Deliver

The capacity of the five producers (and others) to increase their production potential has been questioned by industry observers. Saudi Aramco's claims regarding the reserves it expects to develop were challenged in 2004 in a paper by Matt Simmons, an energy consultant and investment banker (Simmons, 2004). Simmons argued that global energy policy was made on the back of a dangerous assumption: that Saudi Arabia could sustain its spare capacity and continue to increase production. What if it could not? He suggested the kingdom's fields might be in decline and called for Saudi Aramco to make its geological data public. Media attention was galvanized around the ensuing debate, which (very unusually for the media-shy Aramco) took place in public at an American think tank. Dr. Saleri, with other colleagues from the Reservoir Management team at Saudi Aramco, responded to Simmons's claims with a demonstration of the company's forecasts and plans for expanding production. In doing so, they disclosed more geological data than ever before, raising confidence in industry and policy circles that the kingdom's geology was up to the challenge.

Even so, uncertainties persist among governments of importing countries. An enduring concern relates to the stability of Saudi Arabia. Like a number of oil exporters, Iran, Algeria and, of course, Iraq, it presents a potential for political instability. However, it is not within the scope of this book to assess the potential for political instability in these countries.

I can address the second area of concern in the importing countries—whether Saudi Aramco and the Saudi oil ministry can mobilize adequate capital and technology to deliver the promised expansion of oil production capacity in a timely fashion. The Saudi national oil industry's record is strong. Saudi Aramco is a high-capacity NOC. It has demonstrated during the past two years of sustained pressure on oil markets its capacity to increase incremental production in a short lead time when supply showed signs of being insufficient. As noted in Chapter 6, the company has access to sufficient capital for investment in new capacity; and though economic woes for the country loom on the horizon, there are no indications in the foreseeable future that the government will restrict Saudi Aramco's purse. Interviews with E&P executives from the private oil majors have indicated that technologically, Saudi Aramco is on par with industry leaders in its capacity to develop its reservoirs. Those leaders could offer little added value to the development of the kingdom's oil fields.

Interviews at the Kuwait Oil Company suggest that there is a much greater need for foreign investment to assist in expanding capacity than in Saudi Arabia. At present, 80 percent of Kuwaiti production comes from what a Kuwaiti upstream manager called "easy reservoirs, where you could stay home and do nothing and it would get produced." This refers particularly to the Burgan reservoir, the world's second-largest, where a high recovery rate has been achieved from the lower levels by natural water flooding. Thanks to the geology of these fields, Kuwait's average production costs are among the lowest in the world, at under $2/b onshore. The challenge, interviewees felt, is that in order to increase production by 2 million b/d KOC would need to turn to difficult reservoirs, half of which contain heavy oil. Maintaining production capacity targets also means that KOC will need to replace depleted production. A KOC manager explained that Kuwait has an average rate of depletion of five percent per year and that making up for this depletion is difficult. Five percent depletion at the Burgan field means that 100,000 barrels have to be replaced every day. KOC's managers also felt that it was critical to manage water injection in existing reservoirs. The company does not lack adequate financial capacity, however.

The same cannot be said of the National Iranian Oil Company. With insufficient technological and project management capability, it needs IOCs to underwrite its planned increase in production capacity. Iran's crude production of 3.9 million b/d in the third quarter of 2005 averaged slightly below its OPEC quota of 4.1 million b/d. Some observers have warned that NIOC's production capacity could fall to 3 million b/d unless new production comes on stream. The decline rates of aging fields are pushing the oil authorities to step up the development of new fields and to invite participation by foreign companies. NIOC is also looking to offset its decline rates over the next four years by introducing gas-injection schemes in a number of maturing fields. Most investment by IOCs has been concentrated in complex fields where their superior reservoir-management techniques are most needed. Such were the deals on offer in the 1990s. But U.S. pressure on Iran may lead the authorities to offer green fields to foreign majors in order to keep them involved in the country. Iran hopes too that the presence of IOCs will support the development of an export-oriented gas industry.

In Algeria, the need to expand production led the government to open the country's doors to foreign oil companies. Frustrated by the slow pace of the development of resources and by disappointing exploration results, the Ministry of Energy and Mines invited IOCs back to work with Sonatrach. With often superior access to technology, capital and management practices, they have accelerated production in Algeria. As a result of both learning with foreign partners and responding to the pressures of a more competitive operating environment, Sonatrach has increased its technical and managerial competence. Interviews suggested its employees felt that significant improvements to the company's structure and management processes had been initiated in 2001. Also, as pointed out in Chapter 6, Sonatrach is able to gain access to international capital markets and can therefore mobilize sufficient financial resources for its investments. By giving Sonatrach increased independence to pursue a commercial strategy and by inviting foreign investors to assist, under clear terms, in the development of the more challenging hydrocarbon reserves, the Algerian government has increased the company's ability to raise production.

ADNOC is an unusual case. Its regional peers have questioned its ability to deliver production increases, but its foreign partners vouch for its competence. A number of the company's foreign partners stated that Abu Dhabi is able to increase production with immediate effect without even needing to flare gas.[2] The disparate views on the company's ability may be the result of the different type of relations each industry group maintains with ADNOC. Other producers tend to be in contact only with the parent company, but ADNOC's partners work closely with national engineers and other hands-on experts in the operating companies, who are reputed to be highly competent. However, even within the operating companies Abu Dhabi executives have explained that they need to acquire new skills and technology in order to manage large and difficult fields, such as Upper Zakhum. Enhanced recovery in these fields is a particular challenge.

At the parent company level, some foreign partners have commented that ADNOC is laboriously slow and hesitant in taking decisions. One reason for this caution is that the emirate can afford to proceed slowly, as will be made

2. Gas is a by-product of oil production which can be either burned (flared) or captured for use elsewhere in the system (for electricity generation or reinjection into the oil reservoirs, for instance).

Table 8-2. Gas Reserves and Production of Algeria, Iran, Kuwait, Saudi Arabia and the United Arab Emirates, End-2004

Country	*Proved natural gas reserves (trillion cubic meters)*	*Percentage of global proved natural gas reserves*	*Gas production (billion cubic meters)*	*Percentage of global natural gas production*
Algeria	4.55	2.50	82.0	3.0
Iran	27.50	15.30	85.0	3.2
Kuwait	1.57	0.90	9.70	0.4
Saudi Arabia	6.75	3.80	64.0	2.4
UAE	6.06	3.40	45.8	1.7
Total		25.90		10.7

Source: *BP Statistical Review of World Energy 2005.*

clear in "Economic Background," beginning on page 235, which analyzes the future economic needs of the UAE. Some foreign partners told me that Sheikh Khalifa is under pressure from the Bush administration to step up production capacity. Emiratis themselves are loath to discuss such political pressures.

Gas Strategy

Demand for gas is forecast to grow faster than for any other source of energy; consumption is estimated to double by 2030 (IEA, 2004: 129). The Middle East and Algeria hold 43 percent of the world's proved gas reserves. (For the reserves of our five producers, see table 8-2.) However, they supply only 13 percent of its gas consumption (BP, 2004). Historically, this reflects the cost of transporting gas to markets that are adequately supplied by domestic or regional production and also the lower value of gas exports. Nevertheless, with world gas demand expanding and North American and East Asian supplies limited, the development of gas is becoming more appealing. I wanted to assess in interviews the relative importance of gas to crude oil in the producers' strategies. Do the holders of the world's largest gas reserves see themselves as gas producers?

In Saudi Arabia, gas was not mentioned spontaneously in conversations; but when asked about it, respondents said it was very important. I took this as an indication that Saudi Aramco continues to identify itself more as an oil producer than as a gas producer, despite the country's large gas reserves. In fact, the company will be a significant gas producer. Exports of gas are not on the horizon, however; and lower earnings from domestic gas sales mean that Saudi Aramco does not draw as much value from its development of gas as it does from oil. Nonetheless, gas is of high strategic importance to the kingdom to the extent that its development can create opportunities to diversify the economy, for instance by contributing more feedstock to the emerging petrochemicals industry. Gas also meets a growing proportion of domestic electricity needs. Replacing oil by gas in the domestic energy market will mean that more oil is freed for export.

For these reasons, an executive responsible for new business development explained that gas upstream and midstream is a greater priority than refining. Also, Saudi Aramco wanted to develop a gas value chain—from exploration to the development and transport of natural gas to the expansion of the gas-fueled downstream sector—with participation by foreign companies. Foreign investment would help to lessen the enormous capital burden that building a gas chain involves. This is the real test of the Strategic Gas Initiative: it must give Saudi Arabia sufficient natural gas (at a reasonable cost) and gas liquids to meet rapidly expanding demand on its power and water production capacity and from its export-oriented petrochemical industry.

KPC's corporate planning department highlighted the importance of gas, though from a different perspective. It explained that domestic energy demand would grow by seven percent annually and that this increase had prompted a gas initiative. This initiative consists of plans to both produce and import gas so as to meet new demand by 2006. However, there is little non-associated gas in Kuwait, and the central concern has been how to import it from neighboring Qatar, Iran and Iraq. A senior corporate planner explained that Kuwait had engaged in aggressive negotiations with these countries in an effort to import their gas on good terms. Some Kuwaiti industry observers have been critical of the government's limited achievement in this respect.

Strategic perceptions of gas are similar in ADNOC. It is the country's gas supply guarantor, and it faces the challenge of supplying gas to a domestic market that grows by 10 percent annually on average (up to 14 percent per annum in recent years). Gas is needed for power generation, feedstock for the petrochemical industry and reinjection into the oil fields. A number of foreign oil companies would like to invest in the development of the country's vast gas resources (and there are similar untapped ambitions among ADNOC's gas experts as well), but they must realize, as do the ADNOC experts, that the country's leadership does not see the expansion of gas production as a priority. The top leadership has decided to import gas from Qatar's prolific North Field through the Dolphin gas pipeline project rather than to increase its own gas production. But this policy might be revised if these imports cannot meet national demand.[3]

Though Abu Dhabi ranks fifth worldwide in terms of gas reserves, a number of respondents in the company pointed out that its gas reserves are scattered and of various types, a geological fact that makes them expensive to produce and gives Qatari gas a comparative advantage. In many reservoirs, gas would require costly treatment in order to make it marketable. The following exchange with a professional in the Abu Dhabi gas industry reveals a typical mind-set:

> VM: "Also, I imagine that psychologically, you see yourselves more as oil producers than gas producers . . .?"
>
> AD: "Oil is cheap, while gas is expensive to produce. Oil is abundant and there's a higher return on investment."

By contrast, gas ranks high in the strategic and policy thinking of Iran and Algeria, which have relatively less oil potential than Saudi Arabia. Algeria wants to develop as a transport center for gas. It plans to build a pipeline under the Mediterranean to Spain and France and a pipeline from Nigeria to transport its flared gas. Sonatrach is also expanding its upstream holdings internationally, in Spain and Peru, in order to create a web of strategically located gas supplies. In addition, it wants to maintain or increase its market share in the United States (now at 20 percent) and to gain access to the Chinese market. The company's emphasis on the development of gas for export results from a change in government policy that came into effect in 1990. As a corporate planner explained, this change followed from a study prepared by the oil ministry and Sonelgaz in which they assessed 1) expected gas export revenues, 2) forecasts of national energy demand and 3) the potential for new oil discoveries to offset declining reserves. Their projections led to a policy shift emphasizing gas exports.

Gas exports are high on the agenda in Iran too. It wants to become a major gas exporter—this is NIOC's third priority, according to a senior corporate planner. This high strategic priority is linked to the fact that the company can retain revenues from export sales of natural gas and its products, a crucial financing mechanism for this cash-strapped NOC. Company planners would like to develop more valuable gas products, such as condensates.

3. A forecast prepared by one of ADNOC's foreign partners suggested that gas from the Dolphin project would not be able to meet the country's expected demand in 2007.

Pipelines are a high priority too. A plan has been floated since 1993 to build a pipeline to India via Pakistan. A consortium was formed consisting of the Indian Oil Corporation, Gail Oil, the Natural Gas Corporation and NIOC, and the Iranian government has offered to bear 60 percent of the cost of this pipeline (negotiations have been conducted through Iran's Ministry of Foreign Affairs rather than NIOC itself). But various political obstacles have halted progress on most fronts.

Domestic gas needs pose a challenge to export plans too. In Iran, increasing the production of oil depends critically on expanding the production of gas for injection into oil reservoirs. Gas substitution in domestic energy consumption is central to the hydrocarbon strategy of both Iran and Algeria.[4] Increasing domestic gas consumption frees the maximum amount of crude for exports, but gas is also seen now as a valuable commodity for export. Though it still takes second place to crude oil, there is much talk in both countries about the growing global demand for gas and about ambitions to dominate emerging gas markets (Algeria's position as a gas exporter is already established; Iran continues to import gas).

In none of the five producing countries were gas exports perceived as potentially undermining the market share of oil exports. The fuels are not seen as competing with each other. Most respondents in Algeria and Iran, for example, felt that these fuels targeted different markets and that demand for gas would grow.

Integration

When the national oil companies took over from the concessionaires upon nationalization, they held on to the upstream assets but had little or no refining and transport assets. An imbalance resulted between the NOCs, rich in reserves, and the IOCs, which had extensive international downstream facilities and strong marketing networks and skills. Initially, this was not a problem. In the 1970s, the NOCs enjoyed a strong demand for crude oil. This gave them time to learn the mechanics of the core business of upstream oil production. They focused on their domestic markets and sold the bulk of their crude to the

4. Iran's goal is to increase the use of gas in the residential and industrial energy mix by 50 percent, from 55 percent in 2005 to more than 80 percent by 2010. According to information gathered in interviews, gas in Algeria supplies 45 percent of all domestic energy consumption. A Sonatrach planner expected future energy consumption to be "*en profondeur plus qu'en surface*" (deep rather than surface) because the country is already 97 percent electrified. In depth means per capita, while surface means rural, covering territory.

integrated IOCs. However, this segmented market was not conducive to volume coordination between the NOCs and the IOCs. Far from the markets, the NOCs were not attuned to market trends. Also, both sides wanted an advantage over the other, and this led to frequent market imbalances.

Markets were destabilized by geopolitical events as well. Western concern with energy security and the political instability of producers were heightened with the Islamic revolution of 1979 in Iran and the Iran-Iraq war between 1980 and 1988. Consumer states responded with energy efficiency policies in addition to foreign policies aimed at containing the impact of revolutionary and expansionist producers. Meanwhile, buoyed up by strong demand and impelled by the need to replace reserves lost to nationalization, the IOCs developed new sources of oil outside OPEC. When these came on stream in the early 1980s, prices fell. Significant losses in market share, coupled with a period of low prices in 1985–86, led to a sharp decline in oil rent for the OPEC states. Between 1979 and 1985, the supply of oil by OPEC fell dramatically, from 31 million b/d to 16 million b/d. Average prices of $40/b in 1980 fell to $28/b in 1985 (Olorunfemi, 1991).

This uncomfortable new environment led the NOCs to initiate a review of their structure and strategy. They needed to minimize the impact of market instability on their revenues and to maximize the value added on their oil. By investing in downstream facilities in the consuming countries, a number of NOCs spread their risk, secured new outlets for their crude, improved their understanding of the market and consumer needs, added value to their product and hoped to offset any losses in one sector with profits in another (profits in refining often did not materialize, however).

They also sought to develop downstream skills and assets so as to emulate the industrial development pattern of the international oil companies. They integrated their activities downstream, which included refineries, petrochemical production and marketing, and natural gas gathering and export through liquefied natural gas. Their integration effort also included investments in the midstream activities of transport, pipelines and shipping. This challenge of integration and of diversification of activities and geographical focus has been met unequally by the various producers, as shown in figure 8-1, which compares the refining capacity of the five companies under study against their crude oil production capacity.[5]

5. The exceptions are ADNOC and Sonatrach, whose integration has been calculated on the basis of actual production owing to a lack of reliable figures for the capacity of crude oil and condensates. In some cases, joint refining ventures abroad involve long-term crude sales to the joint venture partner.

Figure 8-1. The Five NOCs—A Measure of Integration, 1993–2003

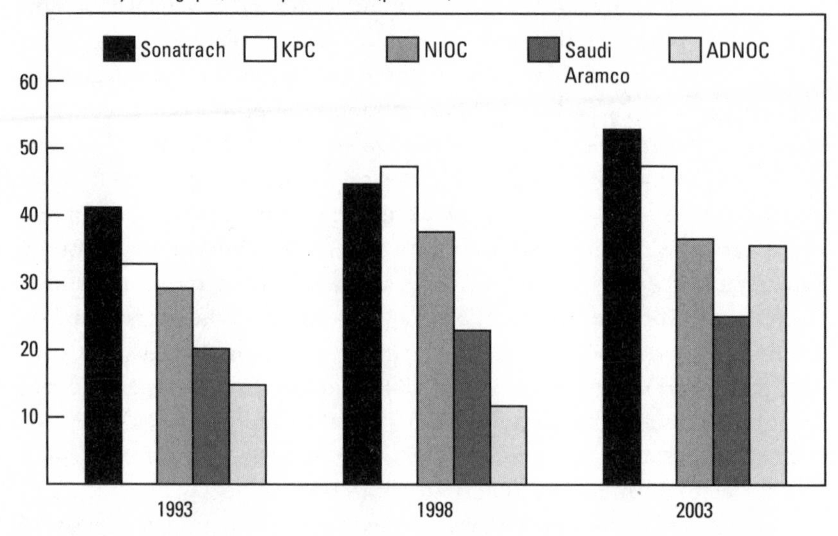

Trends toward integration over the past decade have differed from company to company. Saudi Aramco's integration was partly piecemeal, partly planned. It gained a great deal of new refining capacity when it merged in 1993 with the other Saudi NOC, the Saudi Arabian Marketing and Refining Co. (Samarec). It has also invested significantly in international downstream acquisitions. Before the merger, Saudi Aramco had not been marketing refined products internationally; it had to develop outlets and expertise quickly. Now the company wants to continue integrating, and thus it seeks to develop its capacity and expertise in petrochemicals, transport, international refining and marketing. As part of increasing its capacity in refining, Saudi Aramco will build a new export refinery (400,000 b/d) at Yanbu on the Red Sea coast. In partnership with ExxonMobil it has concluded an agreement with Sinopec to upgrade a refinery in Fujian (taking a 25 percent stake of the expanded 240,000 b/d capacity). And it may also partner with Sinopec to build a 200,000 b/d refinery at Qingdao, due to be commissioned in 2007.

Sonatrach shows increased levels of integration by almost 12 percent in the last ten years (rising to 53 percent in 2003). But this integration has not grown by way of an increase in refining capacity: its five refineries in Algeria appear to have maintained their total capacity of 462,000 b/d over the past 10 years, while its crude and condensate production from its joint venture operations has declined. Meanwhile, total crude oil and condensates volumes accruing to foreign partners more than tripled between 1998 and 2003 (from 12.4 percent to almost 40 percent) in line with more discoveries and increased development efforts.

Conversely, ADNOC showed an initial decline in its level of integration. This is because there were no facilities available to process its new production of condensates between 1996 and 2000. When two specially built condensate-processing facilities began operating in 2000, its level of integration tripled from around 12 percent to 36 percent between 1998 and 2003.

Progress in KPC's integration was disrupted by the Iraqi invasion in 1990 and by a refinery explosion in 2000.[6] Nevertheless, refining capacity has risen since the liberation of Kuwait in 1991, in part owing to acquisitions of European refining outlets. In interviews, KPC managers often said that the company had always sought to maintain a balance between crude sales, at 1.1 million b/d, and crude to refineries, at 0.9 million b/d. KPC's expansion plans include two further refineries for 2010 of a combined capacity of 800,000 b/d.

The measure of integration used for figure 8-1, total refinery throughput as a percentage of crude production, obscures the fact that refined products sold on the domestic market of producing countries are often subsidized and therefore fail to maximize profit for the NOC. In the cases of Sonatrach and KPC, for instance, the figure shows an increasing refining-to-production-capacity ratio, but there has been a reduction in the percentage of oil-based exports refined by each NOC. This is due to increased consumption of refined products at home, including those going into the production of petrochemicals. Domestically, refining capacity is important because it contributes to the NOCs' objective of providing energy to national industry and consumers and thereby to sustaining economic growth. If they do not have sufficient refining capacity to meet domestic demand, they must import oil products, for example gasoline, at market prices, as does Iran. Furthermore, large OPEC producers need to secure dedicated outlets for their daily production. In the case of Saudi Arabia, its domestic refineries take 1.8 million b/d of its crude,[7] a significant relief for Saudi Aramco's marketing efforts.

A second measure of integration provides a useful contrast. Figure 8-2 measures integration as the volume of crude refined for foreign markets in

6. Mina al-Ahmadi refinery was shut down, but it recovered 35 percent of its original 450,000 b/d capacity by 2001–02 (see KPC, *Annual Report 2001–2002*).

7. Saudi refineries, together with Saudi Aramco's 50 percent share of joint ventures in the country, take approximately 1.5 million b/d of Saudi crude (calculated from figures given in Saudi Aramco, *2003 Annual Review*, p. 59).

Figure 8-2. The Five NOCs—Market Export Integration

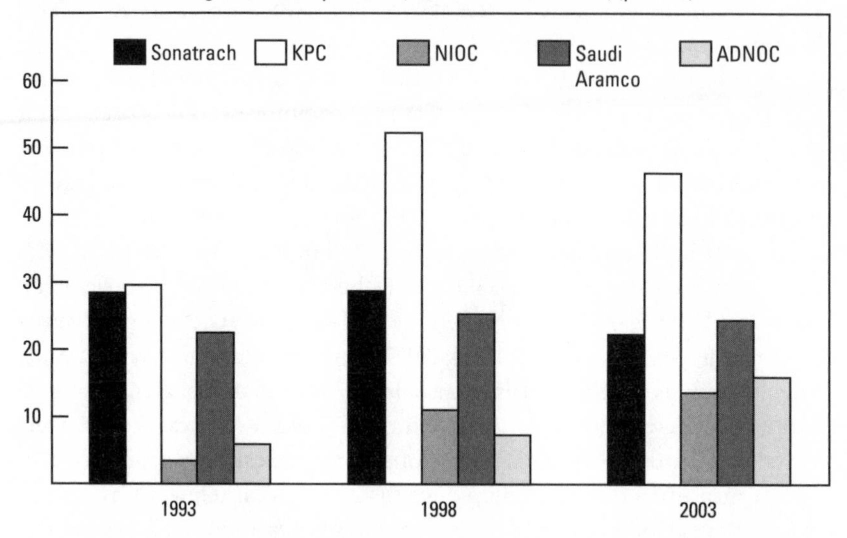

relation to the total volume of crude and its products sold abroad. For KPC and Saudi Aramco, levels of integration are roughly similar in both figures, while Sonatrach and NIOC display much lower levels of integration using the second measure. This is because a greater share of Sonatrach's and NIOC's refined products is consumed domestically. Sonatrach, for example, had the capacity to refine 53 percent of its crude in 2003, yet refined only 22 percent of its crude export volume. NIOC[8] had the capacity to refine 37 percent, yet refined only 14 percent of its crude export volume. Algeria's and Iran's large populations and low cost—and therefore high consumption—of petroleum fuels are the main reason for the disparity. Although NIOC's investments have served to increase its export of oil products over the past 10 years, production has not kept pace with Iran's voracious consumption of heavily subsidized gasoline. Plans are afoot to invest $17 billion in refining capacity between 2005 and 2015, with foreign financing contributing $12 billion of this to increase its refining capacity from around 1.47m b/d to 2.5m b/d. In Abu Dhabi's case, the domestic petrochemicals industry's feedstock needs account for the lower export market integration results.

8. The refining company NIORDC is officially separate from NIOC, but it is considered a part of it for the purposes of this study because NIOC provides it with feedstock at an internal price and the companies are very closely linked.

Saudi Aramco sells on average almost 2 million b/d of refined products (of which half is produced in Saudi Arabia) and 8 million b/d of crude, giving a refining-to-crude production ratio of 1:4. In interviews, I highlighted the contrast with the Kuwaiti oil industry, which has a ratio close to 1:1. An Aramco vice president explained that "We [produce] too much [crude] oil. Our customers want to be independent. [They want our crude] to refine it themselves." The difference in the two countries' refining-to-crude production ratio is also a result of the relative ease with which Saudi Arabia can sell its crude.[9]

Kuwait's high level of integration is tied to the particularities of its crude oil. Kuwaiti crude is "heavy" and "sour," which means that it requires a dedicated refinery to upgrade the residual fuel and to remove the sulfur. As we saw in Chapter 7, this crude is not in demand because more stringent environmental regulations require greater desulfurization. The Kuwait Petroleum Corporation had to build dedicated refineries in order to ensure outlets for its crude. It therefore worked to expand its international refining holdings so as to secure markets for its crude. Owing to the significant share of refined products in its export mix, it is particularly focused on finding markets for those products.

By contrast, Sonatrach has a declining volume of upstream production, and therefore it buys crude produced by its foreign associates in order to feed its domestic refineries. It follows, of course, that the company does not need to secure new refining outlets for its crude.

In addition to providing oil products to meet domestic demand and outlets for crude production, investment in refining has traditionally been seen to provide a degree of security against the volatility of the world oil market. This is because when crude oil prices are low, refiners are more likely to make profits. I asked producers whether they thought that increased integration of their activities through refining helped them to weather bad times and to increase their total profitability. Those interviewed tended to challenge the soundness of this motive for integration. Of course, producers want to make money in refining, but profits are not the only factor that drives integration strategies. Even over several years, refining is not a profitable business for most producers, and it may not be sufficiently efficient to act as a revenue buffer in times of low prices. The lack of profitability of the downstream business has also troubled private international oil companies, despite their greater efficiency.

9. As a downstream executive at KPC explained, Saudi Arabia has five grades of crude, which makes it easier to find outlets for all its crude. Kuwait has one grade of crude.

National producers thus gave other reasons for integrating. Fundamentally, they want to add value to crude (or to "the molecule," as a Saudi official put it) and to be active throughout the transformation process. This is in part a question of pride: they do not want to be simple "extractors" of crude oil; they want to participate in the same industrial activities as their peers in the private sector. As a senior KPC executive put it, "We ask ourselves, do we want to be a raw material producer or add value through the chain?"

Industry experts argue that adding value is a misguided objective and that integration should aim to segregate outlets in such a way as to put a smaller proportion of crude in the open market, a strategy that makes sense only if the refineries are not subtracting value. However, the adding value rationale appears to have its roots in the Middle East's experience of imperialism. The post-colonial worldview frames the national oil company's industrial development in Marxist and Developmentalist terms, along the lines of "Colonial empires exploited our raw materials and sent them back home to be transformed." This means that the sophisticated, value-adding transformation of natural resources escaped the producing country. As an Iranian executive explained, it is a source of pride for the producers to run their own refineries. Now, they want to assume further control over the transformation of raw material to end-product.

It is a sign of a maturing industry that integration efforts in the Middle East are increasingly turning to petrochemicals, which extract by-products such as methane and ethane from oil and gas production and create products ranging from fertilizers to plastics. Most producers are actively seeking to include petrochemicals in their core activities. With the current high oil prices, the petrochemical industry in Europe and America has suffered—but this is an opportunity for the Gulf producers, who have not only access to cheap oil but also vital ethane feedstock. Industry observers expect that by the end of this decade, up to $35 billion will have been channeled into Gulf states' petrochemical projects.[10]

There are several obvious pluses to this from an NOC perspective: the region is open to foreign investment, and petrochemicals can be seen as part of the government's diversification strategy and can create more employment. NOCs have been debating whether petrochemical projects should be given to the private sector, local or foreign. An Iranian Ministry of Petroleum adviser explained that the minister had asked NIOC to keep the petrochemical companies small so that they could be sold later to the private sector. In

10. *Middle East Economic Digest*, 48:45, 5–11 November 2004.

Saudi Arabia, petrochemicals are part of an emerging strategy to attract foreign investment and build on the potential synergies with refining. They will be offered alongside refining assets in order to make the latter more attractive to foreign capital.

Conclusion

The national hydrocarbon industry in the Middle East and Algeria has matured significantly since nationalization. NOCs have been developing their countries' hydrocarbon resources largely independently since then—especially in countries where they were given exclusive access to the mineral domain. However, some NOCs are finding it difficult to achieve new production expansion targets because of technology or capital needs. A general challenge relates to the timing of such investments and concerns regarding the cyclical nature of market prices.

The core business of NOCs has expanded. Initially, they focused on the extraction of crude. But they have learned the refining business and are now engaged in a number of other activities ranging from petrochemicals to sophisticated marketing activities for oil products. Where do they see the limits of their business? Are they trying to emulate international oil companies and integrate their activities internationally? This story is continued in chapter 9.

nine
Going International?

The scope of national oil companies' activities is becoming more international. As we saw in chapter 3, this change is affecting their corporate identity. As a result, there is a great deal of curiosity about, and admiration of, national oil industry leaders such as Petronas that have succeeded in ventures abroad. This Malaysian NOC, like its Chinese and Indian counterparts, is pursuing an aggressive internationalization strategy in an effort to secure new oil reserves. A Petronas executive explained that "Now we are a little bit of a hybrid. We are in more than 30 countries. And 25 percent of our reserves come from outside the [Malaysian] concession. And 20 percent of our revenues are generated from Malaysia." The share and value of their international assets may fall well short of that of the majors, but companies such as Petronas have secured contracts in areas where IOCs have had less success and they are expanding their international portfolios. The emerging trend of upstream internationalization of national oil companies is clearly more threatening to international oil companies than is the NOCs' expansion of their downstream activities, which has been the Middle Eastern NOCs' main area of international activity.

In both downstream and upstream activities, the NOCs' increasing international presence offers advantages: it can provide a much-needed antenna to the world; it increases their exposure to new ideas, technologies and processes; it challenges the NOCs, used to operating without competition, and imposes a concern for cost control. Sonatrach, for one, is expecting a great deal from its internationalization strategy. However, failures are inevitable and difficult to explain to the state. NOCs are indeed traditionally risk-averse. For these and other reasons, the trend is not strong with all NOCs discussed in this book: NIOC, like ADNOC, has focused on the development of national reserves.

NOCs will also have to deal with the increasing penetration of their national market by international players. Internationalization and, conversely, the opening up of domestic markets to competition will be the greatest challenges to NOCs in the future, according to Adnan Shihab Eldin, acting secretary general of OPEC.[1] Moreover, the oil industry's landscape is changing: as IOCs are becoming international energy companies, so NOCs must make the transition to national energy companies. Becoming more involved in the international sphere while increasing the integration of their activities is the current challenge for many NOCs.

At this stage, it is useful to compare the progress that each Middle Eastern NOC has made in internationalizing its activities in various sectors. Each NOC has a very different approach when it comes to making investments abroad, and these are linked to differences in individual market needs, foreign influences and corporate style.

Downstream

Saudi Aramco, the Kuwait Petroleum Corporation and the National Iranian Oil Company have internationalized their downstream activities. (Figure 9-1 shows the extent of their progress in the area of refining.) Sonatrach and ADNOC have no international refining capacity as yet. Saudi Aramco's downstream is growing faster internationally than domestically, and it has more than doubled its international refining capacity since 1993 (from 0.32m b/d to 0.76m b/d through its share in joint ventures). In the early 1990s, it made vigorous efforts to develop downstream joint ventures with local companies in Japan, Korea, Thailand and the Philippines, but with limited success. Renewed efforts led to the acquisition of refining interests in South Korea, the Philippines, Greece and China a few years later. Now China and India are regarded as very promising. As noted in Chapter 8, Saudi Aramco has joined with the Chinese national company Sinopec and Exxon-Mobil in refinery expansion projects and product distribution agreements in

1. Emirates Centre for Strategic Studies and Research, Tenth Energy Conference, Abu Dhabi, 26–27 September 2004.

Figure 9-1. Saudi Aramco, KPC, and NIOC—Internationalization of Refining, 1993–2003

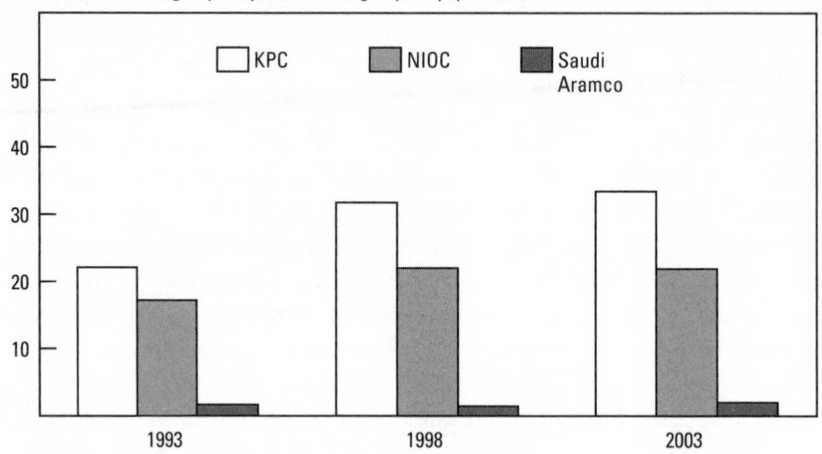

China. The company's present foreign downstream activities began in the United States, where it has a 50 percent stake in the refining and retail chain Motiva. This is a natural progression for a company that inherited an orientation toward the U.S. market and maintained strong connections with the American IOCs after nationalization.

KPC has been more adventurous and willing to take risks. Its push toward diversification and internationalization in the 1980s and the early 1990s was controversial at home. It targeted the European market from scratch in the early 1980s, buying up European refineries and retail chains and developing its own brand (Q8). This strategy provided a secure outlet for Kuwait's heavy crude. However, repeated efforts to establish downstream interests in India, Pakistan, Thailand and Singapore have yielded surprisingly little.

NIOC and ADNOC have been less ambitious in pursuing international ventures, though the Abu Dhabi government owns the International Petroleum Investment Company (IPIC), which creates a market for ADNOC products through its interests in Asian and European refining, retail and petrochemicals. IPIC and ADNOC are separate bodies, but they coordinate their investment plans at the senior management level.[2] NIOC, through its subsidiary Naftiran

2. For example, IPIC supported ADNOC's bid to join with Borealis in building the Bourouge petrochemical complex in Abu Dhabi by buying a 20–25 percent stake in Borealis. When IPIC's interests in refining are factored in, ADNOC–IPIC's joint level of integration

Intertrade Company (NICO), has a 15.38 percent share in the Indian refinery Chennai Petroleum Corporation Ltd. It has made plans to invest in refining in Thailand; but as yet, foreign refining is not a priority.

Sonatrach does not own interests in international refining even though it is active abroad both individually, under the Sonatrach International Holding Company (SIHC), and as part of foreign joint ventures. No precise figures on its internationalization are available but according to the *Oil and Gas Journal*[3] about 10 percent of its total assets were international in 2002. In written comments, the company indicated it is aiming at further international integration through ventures in petrochemicals, trading, gas liquefaction and regasification, electricity generation (and cogeneration) and the development of new pipelines and shipping.

Retail

Securing outlets for refined products is another form of integration. This is pursued by Saudi Aramco, KPC and ADNOC if we include IPIC's investments in some 2,800 service stations internationally. Through Motiva, Saudi Aramco supplies some five percent of the U.S. gasoline retail market. KPC has owned about 5,500 service stations in Europe and Thailand, but it now seems to be scaling down its retail presence abroad.

Upstream

At present, KPC, through its subsidiary KUFPEC, is the only one of the five NOCs to be actively producing oil abroad. Its share of total Kuwaiti production amounts to little more than one percent (around 30,000 boe (barrels of oil equivalent)/d). However, the subsidiary plans to increase this, and it is becoming more independent of KPC. The history of KPC's international forays is inconsistent. Santa Fe Exploration U.K. Ltd., a large upstream acquisition made by KPC, was sold off in 1996 after heavy criticism by the Kuwaiti parliament that it was not profitable enough. Though expansionist thinking ruled in the early 1990s, by 1997 the company was pulling out of a number of its foreign interests. Parliament scrutinizes KPC's international investments intensively (as it did with Santa Fe), perhaps compelling the company to make decisions based on short-term profit or loss rather than on long-term

---

increases from 35.6 percent for ADNOC alone to 49 percent, and the proportion of oil products exported out of total exports increases from 15.9 percent to 35 percent.

3. 16 September 2002.

strategy. The broad trend at KPC has been to divest in the foreign upstream and focus on the downstream and petrochemicals.

Sonatrach has embarked on an ambitious program to acquire international upstream exploration rights with foreign companies. Its aim is to diversify its sources of oil in view of the inevitable decline of the domestic reserve base. In written comments, Sonatrach specified that its strategy is to increase its international reserves to 600 million boe by 2015, with a daily production of approximately 120,000 boe. Since 1999, it has energetically sought participation in oil exploration and production ventures in the Middle East, Libya and West Africa; but as yet, no fields have come into production. The Camisea project in Peru is an important landmark in its international acquisitions, and a senior upstream executive told me that he wants to see this project demonstrate to the world (and the very small network of oil and gas professionals) that an NOC such as Sonatrach has the skills not only to get the job and but also the capacity to carry it out. Sonatrach needs to demonstrate success in its foreign investments in order to gain international credibility.

Saudi Aramco is not active in the upstream internationally and does not appear set to expand its activities in this direction. But, an oil ministry official explained, "We do go outside for every other activity. We have our own fleets . . . refineries (in the Philippines, Korea, Greece, U.S.A., and we are negotiating with China and with India)—[We pursue] only what adds value or gives strategic benefits."

In Iran, the domestic market is large and demanding, and this has kept NIOC focused on its national obligations. However, most respondents in Iran felt that the NOCs that were expanding overseas were successful, commercial companies and that their company would emulate them. NIOC is taking small steps and has not committed itself to an international strategy. As a senior executive noted, "We must be efficient in Iran to be an international company. Internal [constraints] make it difficult to be efficient in our country. We can't wait to reach 95 percent efficiency [to go international]. We are (and will be) a small partner in other projects [in order] to learn, to see how others act, decide, minimize risk. To be operators and have an important role we must be careful about our efficiency." NIOC does have its small international subsidiary, NICO; this has acquired minority stakes in partnerships abroad, including the North Sea area and Azerbaijan.

At this stage, only the senior managements of NIOC (and NICO, naturally) envisage international expansion. As part of plans for the future, two of NICO's arms, Petropars and PetroIran, will merge into one company as

NICO E&P in the hope that as a stronger, consolidated company "it will be taken more seriously by IOCs." As the new company develops its skills and matures, it will become the international upstream subsidiary of NIOC. A senior development executive also indicated that NIOC was planning to significantly develop its international activities in refining, distribution, liquefied natural gas (LNG) and petrochemicals. Others interviewed were not responsive to questions on this subject, presumably because internationalization was not yet on their horizon. However, with the full development of Iran's gas potential, transportation to foreign markets will become a high priority, as it has become for Sonatrach.

Drivers for Internationalization

An adviser on strategy at Sonatrach explained that NOCs cannot remain limited by the confines of their borders. To survive the pressures of competition from other producers and IOCs, to improve relations with consumers and to respond to the challenge of global industry trends such as IOC mergers, NOCs will need to expand internationally and integrate their activities. This strategy is part of their process of industrial maturation. They can no longer be the embodiment of the state, dominating the national economy. The adviser observed that "NOCs have to die to become new actors," just as Elf and Anglo-Iranian died before becoming TOTAL and BP respectively.

In Saudi Arabia, there is no talk of transforming Saudi Aramco, but there are periodic internal challenges that push it to adapt and catch up to industry trends. A senior figure at the oil ministry felt that Saudi Aramco's weakness was its insufficient international presence, unlike "BP, Shell and others, [which] have understood international market forces." Operating only on national territory and with exclusive rights on that territory, a national oil company will not be tested by industry peers. It may be extremely diligent in implementing superior processes for selecting technology, controlling costs and making investment decisions, but without competition (inside or outside the country) who will know whether those processes are as good as possible? Saudi Aramco has already expanded its activities abroad in the downstream and shipping, but the company still has a primarily national focus. Interviews with emerging leaders of the company suggest that it will become more international. This means more investment abroad as well as more foreign investment at home, according to one executive. It also means making more use of its international offices and developing joint research with other companies and universities.

In Kuwait, the internationalization strategy was initiated by the former chairman of KPC, Sheikh Ali Al-Khalifa. As a KUFPEC executive explains: "Sheikh Ali Al-Khalifa [was the company's] visionary for internationalization. It was his project. He wanted the company to be well established, like BP, in all aspects of the business. He wanted also to diversify the asset base of the company, [develop] new markets for the crude, bring in new technology and maximize the returns for the company (the upstream international is very profitable). Kuwait understands oil. It made sense to go do that abroad."

Some NOCs may have ulterior motives for expanding abroad. International activities often allow companies to remove a share of their finances from government scrutiny. PDVSA from Venezuela, for instance, had a number of holdings in the United States that escaped national fiscal obligations.[4] This is an important driver for internationalization when the NOC cannot generate or retain sufficient funds for its capital expenditure. NICO, for instance, NIOC's offshore company, sells and trades its share of crude produced in Iran under the buyback scheme. It does not pay taxes in Iran. The company is designed as a way for NIOC to raise capital without having to rely on budget allocations approved by the government. It also conforms to the legal requirements of the buyback scheme, which states that investing companies must be foreign and generate capital outside Iran. (Their revenues would come from their production, the intention being to help the state avoid paying the companies in case they failed to produce.)

KPC's upstream internationalization strategy too is thought of as a means to generate revenues for the company from crude sales that are beyond the reach of the state, which takes all revenues from the sale of Kuwaiti crude. A corporate planner added, "Upstream, we want to work outside the constraints of OPEC. We want to maximize revenues internationally."

KPI, and especially KUFPEC, escapes government control in other ways as well. KUFPEC can raise its own capital, and does not always have to go back to the central decisionmaking bodies for approval for spending abroad. Also, some Kuwaiti laws do not apply to them. Moreover, these international subsidiaries are in contact only with KPC; they are insulated from much of the government's and parliament's intense scrutiny of national operations. Most notably, they do not receive directives from the government on local content and nationalization of the workforce. KPI has in fact nationalized up to

4. See Juan Carlos Boue, "PDVSA's Refinery Acquisition Program: A Shareholders' Perspective," *MEES*, vol. 47, no. 47, 22 November 2004. Available at *http://www.mees.com/postedarticles/oped/a47n47d01.htm*.

50 percent of its international workforce, but it is not subject to nationalization targets in all its ventures overseas. It has been criticized by parliament for not employing Kuwaitis as pump attendants and in other positions in its international retail operations. But as a KPI executive pointed out, "That would have killed us." The EU's labor restrictions would not have allowed this, and it would not have been commercially viable.

Upstream

NOCs' development of upstream ventures abroad is questioned on the basis that it contributes little to the national interest. Foreign profits are small, because the majority of returns are handed over to other governments in the form of taxes on production. Also, the NOCs which have the appropriate technology and skills and large reserves do not have a clear incentive for internationalizing their upstream business, and their investment capital may be spent more productively on expanding capacity at home. In addition, their skills may not be suitable for foreign ventures: the Middle Eastern NOCs and Sonatrach work onshore, but it is often the case that foreign upstream opportunities are offshore. There are some exceptions, in developing countries with even less technical and managerial resources such as Sudan, but it could be argued that they have much scope for distracting the attention of NOC managements.

Saudi Arabia, with its large reserves and strong national capacity, decided not to invest in the international upstream. Saudis are somewhat resentful of other producers' internationalization, which they perceive as illogical because it undercuts OPEC's price defense policy.

It can be argued that those companies that are seeking skills and whose home production is not so large that their new production outside OPEC would weaken the cartel's capacity to defend the price of oil may have some advantage in internationalizing their upstream activities. This strategy is also relevant to national oil companies in countries with a declining reserve base. This was Petronas's driver for upstream internationalization. But these companies must be financially successful outside or succeed in transferring skills through the company; otherwise, they become a cost-center. Financial success is unlikely in the international upstream until the company acquires the skills needed to compete with the big players outside the protective environment of the low-cost and familiar geology back home.

Companies that have internationalized watch tensely for the success of their international arm. Sonatrach's success has not been rapid, but it is likely to wait out the transition period until its international acquisitions consolidate an advantageous position. Algeria is facing declining crude oil reserves. Finding gas reserves abroad could bring supply closer to the consumer and does not necessarily undercut its markets for nationally produced gas.

KPC does not always have the appetite to take risks with its international upstream subsidiary and to expose itself to parliamentary criticism. After the 1990–91 Gulf war, for instance, Kuwaitis were preoccupied with the challenges of reconstruction, and KPC decided to sell its international upstream activities so as to focus on its core activities at home. However, as a KUFPEC executive explained, it took such a long time to implement this decision that KPC had time to change its mind about KUFPEC's fate. In this period of tergiversation regarding the subsidiary's mandate and future, opportunities for internationalization acquisition were lost while the cash stayed in the bank. KUFPEC's future in the company remains uncertain.

In developing their international upstream activities, NOCs also seek to transfer skills to the national operating company. KUFPEC has sought partnerships with a stake of 30–38 percent because this degree of involvement gives it access to decisionmaking and enables it to learn new skills that it can then transfer to the Kuwait Oil Company. However, the transfer of skills requires good processes for the exchange of information and for joint planning to be established between KOC and KPC. According to KOC and KUFPEC, the two subsidiaries have developed synergies, but one executive suggested that more could be done in this direction if the companies were merged. One sign of efforts to increase collaboration is that KUFPEC will be given an opportunity to take a stake in Project Kuwait: it will work closely with KOC and foreign partners. Furthermore, it was suggested that in order to help KOC to prepare for the next decade, when it will need to develop heavy oil, KUFPEC could be charged with acquiring the needed skills and experience to transfer to it.

The value of this strategy is questioned in the company as a whole because of lack of confidence in KUFPEC's competence. This was also the view of an executive at one of KUFPEC's IOC partners. He felt that its staff were unable to deal with the technological and geological problems that arose during their partnership and that they relied excessively on consultants. In terms of learning and developing best practice internationally, it is worth noting here that KPI's management processes could be more valuable to the parent company than KUFPEC's. It has developed a strong democracy model, with an "open horizontal decision-making structure" in which all employees feel that they have a stake (Tétreault, 1995: 167–68).

The focus on learning skills was also behind NIOC's strategy for setting up NICO and its offshore oil and gas development contractors PetroIran and

Petropars. It expects these companies to learn new skills and technology through partnerships with foreign companies and expects too that as the companies mature and grow, their new competence will enable them to become its upstream development arms. To help in transferring skills to NIOC and to facilitate coordination, NICO, NIOC and other relevant subsidiaries (such as PEDEC) share board members.[5] There is a similar arrangement between KUFPEC, KOC and KPC.

Downstream

With regard to the downstream and petrochemicals, the motivations for internationalization can be summed up as follows:

—Increase proximity to markets (in order to improve understanding of the market)

—Develop closer relationships with customers

—Make a profit

—Add value to crude

—Integrate so as to help insulate revenues against the shock of price swings

—Be active in the whole value chain

—Acquire best practices by working in competition with the international industry (and transfer these skills back home)

—Maintain market share or secure new outlets for national crude

The ranking of these motivations was articulated in a discussion with a senior downstream manager at Saudi Aramco.

> VM: "Why internationalize your business activities?"
>
> SA: "Money first. We want to be active in the whole value chain. We want to improve our understanding of the market. We want to lock in production to the end-users (through refineries and crude outlets). We also want to show our customers and markets that we are with them in the good and in the bad times."
>
> VM: "How do you mean?"
>
> SA: "We will take losses when they do and gain when they do. This will enhance our reputation as a reliable supplier."

5. Three NIOC board members sit on NICO's board, and NICO's chairman sits on the boards of other subsidiaries.

In Kuwait, the emphasis is clearly on profitability—there is none of the altruism implicit on one level in the comment made by the Saudi Aramco executive. KPC and its subsidiaries will sacrifice activities that do not make a profit, even if the activity brings other advantages such as skills transfer and proximity to customers and markets. As a result, the parent company has money in the bank, but it also has less sustained, long-term strategies to develop skills or to take better hold of its market in the future. By contrast, Saudi Aramco will take losses now in a specific activity in order to generate profits decades later in an area of crucial interest.

The Particularities of NOCs' Strategies: Why NOCs Are Not IOCs

The modern oil industry and the oil market itself were created by the first oil companies expanding their activities, initially extracting crude, next taking over transportation and then refining—taking on all the activities through the chain and adding value to the raw material. This integration of activities was driven by the oil majors' desire to control supply, markets and prices. These companies are still very integrated, but they no longer maintain control of their production through to the end-user. Oil is now seen as a fungible commodity, and only a small share (often just 10–20 percent) of the crude feed to the refineries of the private oil majors comes from their own crude production. Crude sources are undifferentiated; the IOCs seek the suitable grade of crude for the best price.

National oil companies appear to be retracing the early footsteps of the IOCs, but they might be on a different path in their integration and, for some, internationalization. The process of their integration is guided by a concern to maintain control, of the product, the price and the destination. For example, Saudi Aramco will sell its crude only to a recognized refiner, as this will ensure that the oil will not be resold at a marginal profit to another supplier. In this, Saudi Aramco and a number of national oil companies have a view of integration that is reminiscent of the early stages of the industry's consolidation, when a handful of companies was able to dominate supply within each of the rings of the chain, from extraction to pipelines to refineries to shipping to retail. Of course, most producers today feel that the price and market control of the 1960s is no longer possible because the chain has been broken by "futures, spot markets, nationalizations and also the Independents intervening in parts of the chain," as a Saudi ministry official explained. But even though big producers still want to control their product and their markets to the extent possible, the process of their integration has taken on particular traits. Middle Eastern NOCs favor long-term relationships with customers, and they do not yet see crude as a fungible commodity. They have specific circumstances that justify their taking a different path.

Getting the Best Price

There is a pervasive sentiment among Middle Eastern producers that they must prevent others from making an additional profit on their crude oil. True, oil is a global commodity, which would imply that there is one global price for each quality (or grade) of crude. But in practice, oil markets are regional: there is stronger demand for crude in Asia than in the West; and Asian consumers compete for Middle Eastern crude, which is closest to their markets. The result of this competition is the "Asian premium," a higher price in Asia.[6] If free market rules governed the global oil market, traders would ship more Middle East crude to Asia, where they could get a better selling price. Because traders are flexible in terms of transport, they can ship the oil bought at Middle Eastern ports to any destination. If substantial amounts could be shipped, they could dissolve the "Asian premium." This would rob the Middle Eastern producers of the marginal profits they make in the Asian markets. But for this reason, the producers exert control over the destination of their crude. They sell it only to customers who have refineries or with whom they have processing agreements and who will not resell the crude. A manager in crude oil sales in Iran explained the precautions that NIOC takes to ensure that new customers respect the end-user clause. NIOC sends a witness on the first cargoes to view the discharging at destination and then to send back a discharge certificate. When the relationship develops, these requirements are no longer needed.

The market in gas is even more regionally segmented. There is no single gas market, and a price formula is set for each country. Strategists at Sonatrach felt that since the removal of the destination clause for European customers, "margins are made at Sonatrach's expense." A French customer, for instance, can resell the gas at a profit to Italy. As a corporate planner put it, "If there is resale, that is a parasitic activity." With LNG, Sonatrach's concern was that it could be resold from Europe to the United States, where the price is higher.

Many producers in the Middle East resent traders, whom they see as parasites threatening to make a profit that ought to come to them and whose

6. Interestingly, an Iranian executive called the "Asian premium" a "Western discount," explaining that the Western markets are well supplied and therefore less dependent.

practices they sometimes see as immoral—"things we wouldn't do because we represent the state," said one marketing specialist in Kuwait. But Kuwait and Iran do sell a small share of their production to traders. It appears that NICO's company, Trade & Finance, based in Lausanne, does most of its trading. Through the buyback contracts in which they are partners, Petropars and PetroIran have entitlements to a share of the production that Trade & Finance trades and sells to end-users for a fee. Sonatrach also sells a portion of its crude production on the international markets through its London trading office, Sonatrach Petroleum Corporation (SPC). Algerians are not shy about their trading activities, contrary to producers in the Gulf. Among the latter, trading is discouraged. The Saudis send clear signals in this respect.

For similar reasons of not wanting to undercut the selling price, the Gulf producers do not favor the spot market. As opposed to term contracts, in which the producer promises a set amount of supply in a rolling monthly contract, spot sales are usually "one-off" and carried out in a competitive market, where the price is set by market forces. According to an international marketing expert at KPC, spot sales require more planning and give consumers better pricing. The GCC is dominated by term contracts, but not in countries where IOCs are investing upstream: the share of spot sales in Oman is 30 percent; it is 30–40 percent in the UAE (though ADNOC's use of them is limited to about three percent) and 60–70 percent in Bahrain. Qatar uses spot sales exclusively. Iran exports only approximately 120,000 b/d via the spot market. Only eight percent of Algeria's traded gas volume is sold by way of spot sales and three percent of its LNG. A Sonatrach manager explained that because there is a surplus of liquefaction capacity compared to that for regasification, the development of the gas supply chain cannot be done through spot prices—otherwise there would be less incentive for the producer to develop LNG exports. He felt that the volume of gas traded on spot markets could increase to 15 percent of the total traded volume. Spot trading is useful for Sonatrach because "It can add value. We use the spot [market] when we can sell the volumes at a good price."

Producers also market to end-users because the proceeds are "the government's revenue and it wants stability," as a KPC marketer explained. Governments' needs are long-term, and therefore the NOCs prefer to build relationships with their customers on that basis. Long-term contracts do not mean that the producers can predict their returns, because prices are benchmarked and change, but they allow a delay, which helps with planning. For gas, national companies in the Middle East and Algeria also seek stable, long-term contracts with customers because demand for gas is not guaranteed to the extent that it is for oil, which is the dedicated fuel for transport. However, the North American market is liberalized, and there are no long-term contracts, only three- to four-year contracts. The European market, which is a hybrid system and still in transition, according to a Sonatrach planner, has only long-term contracts of a minimum of 10–15 years (though the trend is to reduce their length). Some managers at Sonatrach felt that LNG would eventually break these regional markets.

However, in terms of oil sales, governments' need for stability may not be best served by excessive reliance on term contracts. Consultants have conducted studies for the producers on their best strategies for maximizing revenues (and stability) and have generally recommended mixing methods. In the case of Kuwait, they have recommended that KPC put between 15 and 20 percent of its products on the spot market and increase its use of risk management tools by trading 100,000 barrels of crude on futures markets.

Having committed their production to rolling term contracts, many producers have limited flexibility in terms of volumes because most of their oil is "locked in" by these contracts each month. At ADNOC, a marketing specialist was quite frank about the limited impact OPEC directives really have on volumes: "If OPEC cuts production, I am obligated by the entitlement. The volumes are already sold, so when they increase or decrease production in OPEC, it's useless. I can only change a sale two months ahead." There are therefore negative repercussions for price stabilization when OPEC producers are so reticent to engage in spot sales. With so much of the volumes of crude committed through term contracts, there is indeed little flexibility in the system to allow rapid response to OPEC decisions on supply and therefore a reduced capacity to control price fluctuations.

Fungible or Not?

Interviews also tested the five producers' view of their crude as a fungible commodity. With preference given to term contracts and destination clauses, it appears that these producers do not see oil and gas as commodities to be traded and sold in the most lucrative markets. They have a more strategic view of how to supply markets: they seek to keep less profitable customers in order to maintain political relationships and to diversify their demand base. The political backdrop is undeniably there. One of the key directives imposed on Saudi Aramco by the political leadership is to maintain its share of the American market, providing crude at a discount if necessary. This diversification strategy may be informed also by their burning memory of the Asian crisis of 1997: demand withered in what had been the best market. Producers themselves attributed the diversification approach to the fact that they are supplying huge quantities of hydrocarbons every day, implying that they could easily drown a single market with their product. Another factor explaining their preference for stable contracts is the region's commercial culture. In all the countries studied, business is built on long-term relationships with clients. IOCs are appreciated for their respect of "traditional relationships."

This dominant view of crude as a resource with a national flag hoisted on it is being tested by the NOCs' international expansion downstream. Indeed, though refineries were acquired with a view to securing outlets for the national crude, many national oil companies argued behind closed doors that this was not always commercially viable. International refiners at the NOCs felt that their refineries should not be confined to importing the national crude and that they should be able to select the most appropriate crude for the intended product market. In Kuwait, for example, KPI was able to convince KPC and the government in 1986 that its refinery in Italy operated jointly with Agip should take only 60 percent of its crude from Kuwait. This concession was granted because the refinery was a joint venture and allowances had to be made to accommodate the partners. Also, the refinery was not suited for the more sour Kuwaiti crude. Things became more contentious, however, when KPI wanted the other international refineries that were wholly owned by Kuwait to take other crudes. The commercial justification was that Kuwaiti crude is not suited for European refineries. Europe's oil is light and sweet; Kuwaiti crude needs deep conversion and desulfurization in order to meet the standards for European markets. Also, the margins for crude are better in Asia, and thus KPI could argue that Kuwaiti oil could get more value there than in Europe. The opposing argument held that its original mission, like that of other NOCs with international arms, was to secure outlets for Kuwaiti crude. However, it could also be said that the mission had evolved to include the objectives of profitability, diversification and global presence. A KPI executive commented that "Because of that [original] mission of securing outlets for Kuwaiti crude, it was obviously difficult to make a transition where our refinery would accept other crudes. But because of the market realities we couldn't be both profitable and taking only Kuwaiti crude."

The decision to permit the Kuwaiti refineries in Europe to take only 60–65 percent of Kuwaiti crude demonstrates how NOCs are adapting to competition in their external environment. Their expansion to international downstream activities cannot compete with IOCs if they apply the same conditions to those refineries as they would to their national operations. Nor can the NOCs compete with one source of crude, which they need to ship from a distance, while the IOCs swap crudes. A Kuwaiti corporate planner illustrated the difference between the IOCs and the NOCs: "ExxonMobil produces [2.5 million b/d] of crude and it sells 8 million b/d of crude and products. [The NOCs sell more crude than products.] So the IOCs are not as concerned about using their crude for their refineries."

For Saudi Arabia, which produces much more crude oil than products, the natural strategy is for its refineries to take the national crude. Saudi Aramco's refineries abroad are configured for Saudi crude, and oil ministry officials said that their refineries have always taken Saudi crude. This did not appear to be a sensitive question in Riyadh.

Iran is now swapping crude with Caspian suppliers for feedstock to its refineries of sweet crude, which is blended with its sour crude. Iranian refineries are suited to sweet crude, but this grade is generally exported. The country is increasingly producing sour and heavy crude, and so the newer refineries are suited for the new crudes.

The rationale for international integration is changing—from controlling and benefiting from the added value to crude to increasing profitability and competitiveness. There are still managers in the NOCs who feel that the companies should be an instrument of national petroleum policy, if necessary at the expense of profitability. At KPI, an executive explained that the NOCs' need for control is a legacy of their experience of the Seven Sisters and that the memory of those IOCs' way of functioning is still fresh in the Middle East. The region's oil professionals are still digesting the implications of this change. But generally, the NOCs, notably KPC, are responsive to this new rationale. Meanwhile, there is concern about this transition at the political level, and it appears from interviews with KPI that the government and parliament are asking two questions of it: 1) how many jobs are you providing for Kuwaitis? and 2) how many barrels of Kuwaiti crude do you use?

Petronas stands in interesting contrast with the Middle Eastern NOCs in that its managers clearly see oil as fungible. As a result, they view the drivers of their upstream international strategy as distinct from Chinese and Indian state oil companies, which seek to secure sources of crude for their home economies and for which oil is therefore not fungible. At Petronas, new oil production will be sold on international markets at the best available price—not sent back to Malaysia.

VM: "Do you sell the crude you produce internationally to Malaysia first?"

Petronas: "No. We sell it for the best price. Some politicians don't understand. For example, in Malaysia we export our crude and import other crude (like sour crude). Our refineries are very good and can take sour crudes. So we can export our light sweet crude. Otherwise we lose value on our crude."

Hedging or Gambling?

The governments of Iran, Kuwait, Saudi Arabia, Abu Dhabi and Algeria resist the introduction of risk management tools. An executive at KPI explained that there is a concern that hedging by OPEC members would be taken as a signal that "we're throwing in the towel and that we felt we couldn't control prices through OPEC anymore." As a marketing expert in Algeria said, "The greatest tool for price stabilization is OPEC membership. OPEC is the great specialist in hedging." Similarly, an Iranian marketer and many of his counterparts in the other NOCs felt that the market would read too much into their prediction of future prices: "The market is transparent; and if Iran goes to IPE [the International Petroleum Exchange] and starts selling papers at a price, they will think we know something." He also felt that Iran was selling too many barrels per day to hedge. A senior Saudi marketing executive gave a similar view: "We were tempted, but we feel it would disturb the market. We don't know that for sure, but we must be prudent. We may be accused of manipulating the market . . . After all, paper buyers and sellers work on psychology. Markets are not rational. [Really], oil analysts don't know what they're talking about!"

Through NICO's Lausanne-based trading office, Iranians have engaged in some hedging activities. The parent company explains that it is cautious and hedges mostly physical volumes, unlike Western business, which has shifted from hedging physical delivery to speculating on short-term price movements for profit. The Iranian company's caution with hedging is presumably to ensure compliance with Islamic law, which requires investments to be tied to material goods and prohibits speculation. With physical hedging, NICO's office can be assured a buyer is available in the future without writing a contract with a specific customer. It "sells" contracts for 2007 at a set price now, knowing that it will honor those contracts with a delivery of oil. Until then, the contracts may change hands several times, with buyers speculating on the future price of oil. Speculation has been a risky strategy for some. Certain U.S. airlines have recently neared bankruptcy as a result of a shift from pure hedging for future physical needs to speculating for profit. They had purchased futures contracts for physical delivery of oil so that they would know their costs and could price tickets. However, they sold their futures contracts for a profit, and now are unprotected from rising oil prices.

For the producer, physical hedging carries risks too. NOCs do not like to gamble with the possible returns for the state: if the market price ends up exceeding the futures' price expectations, the lost returns are hard to explain. The burden of this accountability is also felt by the subsidiaries, which have to explain any revenue shortfall to the parent company. A marketing director in the Emirates felt that the risk could be justified only if the state needed to secure a minimal amount of income, for its debt service, for example. Moreover, futures involve an upfront cost. As a Kuwaiti executive explained, "The problem is that futures costs are taken from my margin. If I have a $50 million contract, am I willing to pay $2 million at the beginning to avoid having $5 million in losses?"

The habit formed over decades of selling crude by term contracts has also made many marketing people in the NOCs resistant to the introduction of new selling techniques in which they lack expertise. A majority of those interviewed were not very familiar with futures, and, inexplicably, some in Sonatrach grew agitated when asked about it. The NOCs would need to turn to others to provide this service. They have been approached by the majors, banks, the New York Mercantile Exchange (Nymex) and trading companies, all anxious to explain why it is to their advantage. But many felt that hedging tools are more suited to traders.

Conclusion

If national oil companies are following the private oil majors on the path they have taken since their days as the Seven Sisters, they have a long way to go before they catch up. However, this chapter has shown that though some NOCs wish to emulate the private companies in their international reach, they are not on the same path as the private majors. This is because the NOCs have different assets that shape a distinct set of objectives for internationalization. While the majors comb the world for new upstream assets, the Middle East NOCs have large reserves at home and few seek to develop their upstream activities abroad. They produce large volumes of crude and therefore look to international downstream activities as outlets for their production. They also face constraints, some related to their status as a national company, which limit their commercial behavior. They may sell oil at a discount to support diplomatic relationships. They are also conservative with new marketing techniques because governments depend on them to generate revenues. This cautious attitude is also shaped by their culture and marketing habits formed over time.

ten

Partnerships

As we saw in the previous chapter, new trends are shaping the oil and gas industry, notably an increased blurring of differences between NOCs and IOCs. Public ownership is also becoming an elastic concept: NOCs such as the Norwegian Statoil and the Brazilian Petrobras are partially privatized but maintain a majority government stake. Nor can "national oil companies" be confined any longer to their national borders. The political frontier niche is getting crowded: in politically unstable countries IOCs are encountering more competition from developing world NOCs, notably from the Chinese groups Sinopec and CNPC, willing to take bigger political risks and lower returns. High-capacity NOCs such as Petronas from Malaysia are challenging the IOCs on their "territory" of high political risk ventures while Petrobras is tackling exploration in the deep and the ultra-deep acreage with technology that had previously been the reserve of IOCs. Most NOCs no longer limit their activities to producing and selling crude. A good example of this change is KPC, which is highly integrated.

These new NOC strategies present opportunities for different forms of partnership with other companies. There are interesting avenues forward that go

This chapter is based on "National and International Oil Companies: Existing and Emerging Partnerships," a paper given by the author at the Emirates Centre for Strategic Studies and Research's Tenth Energy Conference in Abu Dhabi on 26–27 September 2004 (ECSSR publication forthcoming, 2006).

beyond the conventional arrangements between IOCs and NOCs, such as production sharing agreements (PSAs). The conventional PSAs, which offer IOCs access to equity, have become an essential component of the global oil industry because they are, in addition to mergers and acquisitions, the principal means by which private oil companies increase their reserves base. These "booked reserves" are represented in the IOCs' balance sheets; they are an important criterion by which those companies' future stock value is measured by the financial markets. Equity, together with acceptable rates of return, incentives for enhanced recovery and opportunities for repeatable investments, are the terms of investment most important to international oil companies.

However, these terms are not on offer in many of the large reserve-holding countries of the world. As indicated in chapter 2, Iran, Saudi Arabia and Kuwait will not give equity access to their reserves. Further, they will seek to maintain control over the management of their resources and to maximize the "government take." Agreement can more easily be reached on returns, but present investment models are not sufficiently flexible to address other producer concerns. A crucial problem with present models of partnership is that IOCs would control the management of reservoirs without sufficient incentives for optimizing the development of the host country's resources. Moreover, conventional partnership models do not make full use of the emerging trends transforming the industry to develop new strategic relationships.

A wider range of strategic partnerships must be developed because one model will not fit every case. Each national oil company is unique and has its own needs, constraints and assets, as table 10-1 on the five NOCs under study shows.

Chapter 8 described the foreign investment needs of the five producers, of which several require assistance in delivering on their ambitious plans to increase production capacity in oil and gas. IOCs, for their part, need to replace every year the reserves they produce (and to add more on top of that) and must seek new areas to invest in. As a result, there is much debate in the industry about how to increase foreign investment in the Middle East. Indeed, the theme of partnerships between national and international oil companies finds its way onto the agenda of a number of industry conferences. At these events, both sides profess their desire to find ways to work together. In private, however, both groups express frustration over the other's expectations and the difficulties of finding common ground. IOCs' frustration with previous joint ventures and NOCs' increased confidence in their ability have led many on both sides to question the need to work with the other—at least under the terms currently discussed.

When asked in interviews how they would define the perfect deal with IOCs, NOC managers across the region described it as a win-win deal, and surprisingly, they highlighted the needs of the IOCs. "It has to be a win-win project. We have political risk. There's the crisis with the Atomic [Energy] Agency, the U.S. embargo. [As an incentive] we can put a long-term clause in our contracts." "[It would be a] deal where we are both happy, both winners. It's bad when a deal favors one over another. A balance is necessary—like in a marriage." Also, there was almost a consensus among the NOCs that the perfect deal would be based on technical grounds, not politics, and many managers thought that the deal should give acceptable rates of return to both parties. "We must pay them right, so they're happy. We shouldn't worry about the contract terms, but we should focus on the resource management."

However, the terms on offer from several Gulf producers do not meet the IOCs' investment criteria. For this reason, producers such as Iran and Saudi Arabia were unable to attract bids from a large number of big players. A Saudi Aramco executive explained that the company was disappointed that the Strategic Gas Initiative did not attract more investors: "We wanted more players. It's too bad ConocoPhillips opted out of the gas initiative. It's a good company. [But they explained that] the bid didn't have the expected returns." To bring the two sides' expectations about returns closer together, producers would like to see the oil majors increase their oil price assumptions to match OPEC's. The majors assumed $18/b for planning in the 1990s versus a $20/b OPEC target, so presumably their price assumption was of $22–23/b in 2004 versus OPEC's $25/b (in September 2004 the OPEC basket price averaged $40).[1] Nevertheless, a number of IOCs do accept less than satisfactory terms of investment for small-scale or short-term investments in order to build relationships that they hope will give them access to reserves and better returns in the future.

This chapter will examine how various investment models can enable both parties to align their interests while responding strategically to emerging industry trends.

Provision of Services

NOCs are under pressure to invest so as to control depletion rates. IOCs will increasingly be called upon by the major reserve holders in the Middle East to assist them in managing their reservoirs and offsetting the decline of their

1. *Oil and Gas Abacus,* Deutsche Bank, 16 September 2004.

Table 10-1. Assets, Needs and Constraints of the Five NOCs

NOC	*Assets*	*Needs*	*Constraints*
ADNOC	High ratio of oil and gas reserves to production and population; cooperative relations with foreign partners; management processes.	Develop HR skills, capacity to manage large projects, ownership of technology; investment in difficult reservoirs; marketing capacity for products; investment in gas (for re-injection).	Political reluctance to develop gas for export; reliance on consultants; local employment quotas.
KPC	Efficient refining, retail and petrochemicals business; international marketing skills; large oil reserves.	Experience of technology; management practices; clarity of relations with government; employment of nationals.	Parliamentary opposition to foreign direct investment (FDI); heavy/sour grade of crude; small domestic market; bureaucratic internal processes; local employment quotas.
NIOC	Large oil and gas reserves; experience with carbonate reservoirs; local capacity in private service companies; geography.	Capital; clarity of relations with government; technology; management practices; investment in refining, exploration and development; employment of nationals; marketing skills for gas.	Sanctions; parliamentary opposition to FDI; domestic energy subsidies; bureaucratic internal processes.
Saudi Aramco	Efficiency; large oil and gas reserves; multiple grades of crude; long-term strategic view; investment in technology; human resources.	Capital for refining and petrochemicals; promote local economy; ownership of technology; outlets for its crude (international refineries); develop gas value chain.	Upstream oil closed; political sensitivity; domestic energy pricing; future rent needs of government.
Sonatrach	Oil and gas reserves; LNG expertise; capitalization on NOC status abroad; geography; transparent accounting.	Management practices; access to distant and new markets; new oil and gas reserves; investment in technology.	Heavy labor costs; bureaucratic internal processes.

Source: Author's interviews.

mature fields. However, IOCs may not want this type of work. Indeed, the current low level of IOC involvement in the Middle East is at least partially attributable to the reluctance of the majors to commit their skilled personnel to this type of activity that brings lower returns. Projects in their own mature areas already present significant technological challenges and high costs just to maintain production levels. The majors seek opportunities to explore and develop new areas to replace the reserves they are producing in the areas traditionally open to them.

But should they decide to work as contractors for NOCs or in partnership with them to develop existing fields, the IOCs would need to engage in a different kind of relationship with the NOCs and their governments. Their activities would not include wildcatting—the risky exploration game; and with less risk, the IOCs must be prepared to accept lower returns. As service providers, they would need to change their mind-set, their corporate culture—and to work less independently. They must recognize that producing countries want to control their resources. Producers want assistance, but on their own terms. Companies that recognize this are most likely to have successful relations with the producers. An NIOC manager explained the producers' perspective:

> Service companies can provide services often at a better cost than IOCs. This is also true with Iranian service companies. They can do exploration services, seismic, drilling, tankers. . . . There are many, many alternatives to IOCs. IOCs have to bring down their expectations and have balanced, equal relations with us. Unfortunately, they are still stuck where they were 50 years ago—they still want PSAs. We decided to propose another formula [the buyback], which has been able to [attract] $40 billion over the last six years. IOCs must change their views and appreciate local capabilities, have balanced relations and reduce their expectations.

In this sense, the problem with the service-provider investment model is not so much one of access as of the conditions of access. If IOCs reassess their expectations, repackage their offers and do sit down to negotiate as service providers, countries such as Iran may open their doors wider. At present, some of the activities of IOCs fit into this approach. In some countries, this occurs in haphazard forms, and there some IOCs may hope (some are bigger believers than others) that they will someday be invited to explore and develop new fields "turnkey."

The service-provider model could in practice result in increasing friction with service companies. The package IOCs offer would need to be defined in relation to the services already provided by companies such as Schlumberger and Halliburton. As it stands, large integrated projects are the domain of IOCs, except where a country is closed to foreign capital investment in the hydrocarbon sector. In this case, the service companies step in. Closed markets such as Mexico and Iran, where foreign companies cannot lift barrels or market the oil produced, have been the domain of service companies, which operate within strict parameters. A recent and interesting development is the new range of services that the IOCs are developing to cater to the producers: financial management tools (hedging, futures), technical consulting, systems consulting, and so on. Here again, IOCs will clash with other service providers such as banks and consulting groups that already offer these services.

On the producers' side, there are also choices to be made. If controlling hydrocarbon resources and maximizing their life span is truly as important to the producers as they say it is, then they will have to propose contract terms that provide incentives for optimizing their development over the long term.

There are, in addition, broader issues to be resolved regarding the role of the national oil companies in the service-provider model. This model is drawing increasing interest from the producing countries because it allows the NOCs to hire companies to provide specific technology and services and thereby fill any capacity gaps they might have. With this model, NOCs can become customers who manage contractors. There is no danger here if the NOC has strong capacity and is able to control the technological and cost terms of the contract. It will have access to the best technology at a lower cost. However, this model is not without risks: NOCs may lose competencies, as their skills will not be challenged by experience. National oil companies that only form partnerships with IOCs and engage in no independent activity of their own end up being sleeping partners even if on paper they have a controlling majority or appoint the senior executives (Luciani, 2004: 17). It therefore becomes more difficult for the NOC to hone the skills necessary to manage the development of its resources. The service-provider model may thus be most beneficial if it is applied to mature fields and complemented by other investment frameworks in other areas.

NOCs' growing dependence on consultants raises a similar concern about maintaining capacity. Managers delegate a number of technical tasks to consultants, in some cases even asking them to read contracts that the NOC must sign. This tends to be a problem in parent companies, which sometimes do not have in-house personnel with the relevant skills—these highly skilled and/or experienced staff are often needed most in the operating companies. The implied dangers for effective decisionmaking were highlighted in interviews with ADNOC's foreign partners. In general, decisionmakers should be encouraged to rely on their own judgment and on the expertise available in their own team. If they distrust the soundness of this expertise, then strenuous efforts should be made to improve the team's quality. Companies could gain from integrating the required talent within their personnel, thereby encouraging them to be part of the organization and to share responsibility for its failures and its successes.

IOC–NOC Partnership to Advance Sustainable Development in the Host Country

The sustainable development of producer states involves maximizing the benefits for their economies today and optimizing the development of resources for the benefit of future generations. Many producers face serious development challenges, and governments count on the hydrocarbon sector to provide more than oil rent. Many of the Middle East's oil professionals felt that IOCs care about finding mutually satisfying agreements but that they are not as concerned about the long-term prosperity of the partner country. Former Venezuelan Minister of Energy and Mines Alirio Parra explained that there is a difference between revenue maximization, which is a short-term concern, and value creation, which is a long-term activity. NOCs, I gathered from interviews, are driven by value creation; IOCs (and to a certain extent governments) tend to be motivated by revenue maximization. These different drivers make it more difficult to align the two parties' interests in a partnership.

Furthermore, there was some discomfort within NOCs about the IOCs' corporate social responsibility (CSR) programs. Some NOC managers gave contradictory views about what the IOCs' responsibility should be in this respect. Such programs challenge the government's prerogative to provide services; the IOCs get the credit and, in the end, are compensated for the cost; the NOC is challenged on its home turf. States pay for the service but are spared the effort of providing it for the population. One executive voiced this ambivalence: "You know that IOCs bring up their [required] rate of return with their spending on charity programs? As I told you before, there is no free lunch."

Many oil professionals felt that IOCs develop expensive programs that NOCs and states could carry out more cheaply. They would prefer that the funds designated for social work be given to the state to spend. They believe that this money is ultimately taken from the returns to the state through the contract terms. Also, many NOC managers considered that IOCs, unlike national institutions, do not understand domestic needs. There is great pride in many NOCs for having hitherto responded to the needs of the nation. But as the state's capacity to provide services to the population grows in the Middle East, NOCs are called upon less to build infrastructure or to provide social programs. As pointed out in chapter 6, a new trend for NOCs is the development of more classic private company-type social programs, which seek to ensure adequate development of the producing region at a lower cost while improving the public image of the NOC.

In this context, if private companies' CSR programs are to bring greater prosperity to the oil-producing states while overcoming the suspicion of NOCs and respecting the boundaries of state prerogatives, they must be coordinated with existing programs handled by the relevant ministries and those put in place by the NOCs. IOCs should support existing state programs with funds and with whatever knowledge and experience of managing development projects they can bring.

In addition to providing benefits to the community, a partnership model might involve the NOCs gaining new competence from the joint development and application of technology. An Iranian professional commented on the potential mutual gain: "NOCs and IOCs need each other. The NOCs offer access to reserves while IOCs offer access to capital, expertise (the latest EOR [enhanced oil recovery], for example), management skills—this is a key thing missing in NOCs. We have too much bureaucracy, which brings bad management, delays . . . all these things disappear with an IOC. For example, if we order spare parts, the request goes to committee after committee. The IOC just gets it . . . And research. Iran doesn't have research centers, except the one inaugurated by NIOC . . . it's a shame!"

In practice, however, it is difficult to transfer skills and technology. Not all NOCs are well equipped to take on new skills, because of weaker management practices and a specific corporate culture. One manager felt that NIOC had been unsuccessful in developing "deep relations" with foreign companies and that interest was lacking for real technology transfer. Some respondents in Kuwait said that KOC appeared to be resistant to technological transfer. Some felt that the company was too risk-averse to accept significant technological changes to its operations. Conversely, IOCs face a challenge in gaining access to a country with a high-capacity NOC, like Aramco. Whether they worked for a high- or low-capacity NOC, a number of national oil experts exhibited residual resentment of foreign oil companies, an inheritance of the time of the consortia.

> As an Iranian who went through nationalization, revolution . . . my view of oil [is very shaped by those events]. I worked in the consortium and was astounded to find a "no Persian-speaking" sign in the managers' mess. I saw that people were accomplishing their private business in the back field and I asked them why they were doing that. "You should use the toilets!" They said there were no toilets. There was no loo for workmen, who of course were all Iranian. So now, when outsiders say that "they will teach me something" . . . .

The present terms of investment are not well designed to facilitate joint development and application of technology. The buybacks in Iran or the joint study agreements in Kuwait have disappointed the partners in this respect. In the buybacks, the IOCs work with Iranian companies and thus do transfer technology; but they are not contractually responsible for teaching skills and technology. KPC has joint-study agreements with IOCs through which they provide analyses of specific themes, such as gas supply or geophysical acquisition, and work in a team with KPC. There is skills transfer in Kuwait, though it appears to be limited, according to an IOC executive. In Sonatrach, the mood is different: the Algerian company appears to be very keen to learn from the IOCs. This is perhaps because it has been put to the test by exposure to competition in its domestic upstream business.

There is no easy formula for a partnership that enables NOCs to control the development of their states' resources while they learn the skills of the IOCs. Terms can be found, but the will must be there too. The crucial issue for NOCs is trust, and the lack of it is a serious obstacle to developing IOC–NOC partnerships. This legacy will need to be overcome in order to meet the industry's investment challenge over the next 10–20 years. New contractual arrangements may be needed to define new types of relationships between producers and IOCs. Breaking with the legacy of the past, especially in Iran, must include rethinking equity-based contracts. For this, private companies must be evaluated on more than their reserves by shareholders and financial analysts. Building partnerships will also require some good old human skills, and can be built on chemistry between people and their relationships.

On the IOC side, cultural sensitivity and good listening are required. IOCs should not underestimate the knowledge of NOCs. The Middle Eastern

NOCs have, after all, kept their industry running with little help from the IOCs for the past 30 years—the Iraqi oil professionals who kept the crude flowing during 12 years of UN sanctions give a dramatic illustration of ingenious self-reliance. That being said, IOCs need to know who in the producer states decides, who regulates, what committees they must refer to and who sits on those committees. They also need a clear legal framework in which to operate, which they do not find often enough. IOCs also want access to a value chain so that they can give the state maximum benefits throughout the chain. They are, after all, integrated companies.

NOC–NOC Partnership

In an effort to identify future trends for NOCs, I inquired in interviews about the potential for NOC–NOC partnerships and collaborations. Are NOCs in the Middle East interested in the assets that their counterparts bring to the table? I found that Middle East NOCs generally distrusted other NOCs in the region almost as much as they do IOCs—though for different reasons. In the case of high-capacity NOCs such as Saudi Aramco, distrust is tied to disdain for the capacity of their national counterparts. For medium-capacity NOCs, there is essentially a lack of interest in other NOCs' assets. For now, Middle Eastern NOCs prefer to deal with the best available investors, and they are the super-majors and their traditional partners. An Iranian and a Saudi commented in separate interviews on the prospects for regional cooperation:

> Iranian: "They [the Gulf Arab producers] are uncomfortable in the region and don't cooperate well. We have a joint field with Saudi Arabia and Qatar (South Pars). They are not even ready to share information on the field for better management of the reservoir! Arab members of OPEC have never agreed to an Iranian as the Secretary-General. So we never agreed to an Arab Secretary-General!! It's very difficult!"[2]
>
> Saudi: "I don't see a lot of potential in regional cooperation between NOCs because Saudi Aramco is so far ahead that what would we gain from them? We tried to cooperate regionally and have been disappointed."

Discussing this question with young professionals, I found that most were, as one young Iranian engineer put it, "100 percent for" NOC–NOC partner-

2. Tensions between Saudi Arabia and Iran arose during the Iran-Iraq war and worsened after Iranian pilgrims were killed in clashes with Saudi security forces in 1987.

ships. But when asked with which NOCs they would like to work, they explained that they would like to work with the company best suited for a specific project, irrespective of whether it is an NOC or an IOC. My impression was that when thinking politically, they appear to prefer to work with NOCs (not necessarily from their region) but that when thinking commercially, they prefer IOCs. However, this is likely to change in favor of NOCs in the future. Investment from foreign NOCs is politically more palatable in the region, and this may help some producers open up their upstream sector. But positive attitudes toward NOC investment are also likely to develop for commercial reasons. A growing number of upstream deals are being signed by internationalizing NOCs from the developing world. They are investing in countries whose terms the IOCs are reluctant to accept; and though they may not be the oil producers' first choice in terms of capacity, it is likely that in time they will gain a comparative advantage as their experience and credibility in the Middle East increase.

For instance, major oil companies demonstrated less interest in the last round of Saudi gas tenders than did Chinese and Russian investors, who stepped in with lower requirements for rate of return and important political and strategic drivers to motivate them. On the one hand, this is likely to worry Saudi Aramco, which has so far been convinced of the attractiveness of the kingdom's gas resources to foreign investors. The Chinese and Russian companies do not promise the highest investment standards and technology. On the other hand, they are quickly honing their skills and may be able to impress Saudi Aramco and Middle Eastern producers more generally. On a political level, these new investors diversify international sources of support for the kingdom. Iran's view of the investors courting its hydrocarbon resources is quite similar, and the government has sought to diversify its sources of support by giving Asian and European countries a stake in the country's stability.

A manager at the Russian company Gazexport commented that Chinese, South Korean, Malaysian and Japanese oil and gas companies are offering the Middle Eastern producers more competitive investment bids than the IOCs. The Chinese are especially competitive, because they develop cheaper technical solutions. The majors themselves are outsourcing much work to service companies in China and Malaysia. Petronas is of particular interest to most observers in NOCs as it quickly develops its international presence. As the Gazexport executive observed, "They've got expertise (learnt from majors), government support, cultural awareness—a country with a 30–30–30 population mix ought to know how to deal with cultural differences." This view is echoed by other managers in the Middle East: "These companies are up and coming, especially Chinese companies. They have a rapidly increasing role in these fields—Petronas especially. Petronas works a lot with European companies. The Indians also."

NOC–NOC partnerships are also being pursued on another front. Sonatrach, Statoil, Petrobras and Saudi Aramco are active participants and organizers of the NOC Forum, which for three years has brought CEOs of national oil companies around a table annually to share ideas on how to develop NOCs' core competences. This underlines the emerging trend in the oil industry of the growth of links between NOCs; they are beginning to look outside their borders and develop an interest in the experience of other producers.

Sonatrach is particularly keen to develop new ties. It has no strategic alliances with the Gulf producers, but it is turning its attention to Africa, notably to projects such as the African Energy Commission. It also has a plan, noted in chapter 8, to transport and provide an outlet for Nigeria's associated gas, which has so far been flared, despite the country's domestic energy needs. This project would fulfill a number of objectives for the Algerian industry: it would advance the plan to make Algeria a gas hub; address a serious pollution problem in Nigeria, which would benefit Algeria's image; and bring energy not only to Nigerians but also to Niger and Mali and to the underdeveloped, southern part of Algeria. The project would also contribute better infrastructure to this region of Algeria.

As pointed out in chapter 7, producers are seeking more and more to develop new technologies with partners. Their motives are twofold: by being involved in the development of new technologies, they can gain a better understanding of their applications, and they get a stake in the ownership of the product. This could be an innovative means for NOCs (and IOCs) to forge close alliances with other producers and build trust. This objective was expressed by ADNOC and KOC, which want to develop reservoir management technologies, and also by Saudi Aramco, which would like to invest in R&D in hydrogen (fossil fuels for hydrogen production).

IOC–NOC–Private Local Company Joint Ventures

Another innovative partnership model would have an IOC join with an NOC and a local private company. Such partnerships have been successful in the petrochemical and refining sectors, and could be developed for upstream and midstream development. A model for such partnerships is the Equate joint venture, successfully developed in Kuwait. PIC, the petrochemicals subsidiary of KPC, joined forces with Union Carbide (since bought by Dow Chemical Company) and local private companies, which took a 10 percent share. Equate is independent from the state and the NOC and functions like any other commercial entity. Its international partners have given the company efficient, Western-style management practices and technical expertise. Some of Equate's good practices have been applied in the PIC subsidiary—it has notably changed its management, IT and HSE processes.

However, the private sector does not necessarily have much appetite for investment in downstream activities. In Kuwait, for instance, Mary-Ann Tétreault has noted the "short breath" of the merchant families, which sell shares back to the government as soon as losses appear—this occurred with private sector participation in the refining subsidiary KNPC in the late 1970s (Tétreault, 2003: 86).

As for the application of this model to the upstream, there is the question of whether states in the Middle East would welcome private domestic equity any more than foreign oil companies. The participation of the private sector in the region's hydrocarbon industry has historically been limited. In Kuwait, the long-standing pact between the ruling Al-Sabah family and the merchant class is that the oil industry is state-controlled and the Al-Sabahs agree to stay out of Kuwaiti business. The merchants have access to lucrative contracts, preferential monopolies and dealerships and receive their share of the oil revenue windfall (Crystal, 1990: 75). It is unlikely that states in the Middle East would relinquish control over the majority of their hydrocarbon resources, because they draw power from their monopoly of the generation and distribution of oil revenues. But some minor hydrocarbon assets could be shared on the lines of an NOC–IOC–local private company partnership, with potential gains for all partners. From the government's perspective, such a deal would reinforce national policies by giving new opportunities to the private sector. At the same time, it would maintain control over the development of oil through the NOC's involvement and perhaps conceal from public opinion the role of foreign interests in the investment. From the NOC's perspective, it offers possibilities of forging new alliances with the domestic private sector. Alliances with foreign partners may, in addition, serve as leverage for the NOC vis-à-vis the state.

Joint Ventures Abroad

Joint ventures between IOCs and NOCs in a third country offer neutral territory in which to develop relations. Here the NOC would learn management practices, the application of technology and project management from industry leaders, provided it had a large enough stake to get a seat on the board. The NOC would also be in a more receptive mode, being challenged by a new environment, and not intent on protecting national sovereignty. The NOC and the IOC partners could develop personal and institutional relationships through these ventures, which would help to overcome issues of trust on the part of NOCs. Through a demonstration of NOCs' assets and skills, IOCs could learn the value of their partners. Such has been, to a certain extent, the experience of Petrobras. The Brazilian NOC sought to pool resources with IOCs through international ventures. This allowed the partners, according to one company executive, to meld together their experience and technology.

Though joint ventures in a third country can help to build a relationship between IOC and NOC that facilitates future deals, they cannot realistically be considered a direct avenue to investment in the partner NOC's country. Nor can the IOC be expected to risk its reputation on an alliance with a weak partner. Joint ventures abroad must be beneficial to IOC partners regardless of potential investment gains in the NOC partner's home territory—joint ventures do not survive for long when they are based on hypothetical gains. As a result, the lower-capacity NOC will find it difficult to attract an IOC into a partnership abroad. International private companies will seek out partners who bring value to the table.

Middle Eastern NOCs bring more assets to downstream joint ventures than to upstream ones, where the operator's responsibility is great and a strong track record in applying the right technology and controlling costs is crucial. Because NOCs have less experience with controlling costs in large projects and new geological conditions, they have a great deal to learn from IOCs in upstream joint ventures abroad. In terms of international downstream ventures, NOCs such as KPC, Sonatrach and Saudi Aramco have already integrated their activities significantly and are active in refining; but they can still learn from the IOCs' management practices, HSE standards and marketing skills (specifically their experience of retail in a competitive market). As a sign of this emerging trend, KPI has signed memoranda of understanding with Shell and BP to explore downstream business opportunities worldwide (and specifically in China with BP).

As indicated in chapter 3, Middle Eastern NOCs are only beginning to identify and capitalize on their assets. By contrast, Sonatrach is quite deliberate in capitalizing on its experience of dealing with the social environment when approaching host countries—some of its peers recognized this as an Algerian advantage. Some Algerian oil professionals felt that their approach to foreign ventures was based on respect for other cultures. They explained that in some cases, Sonatrach had more to offer the host countries than did the majors because of its capacity to listen, its common values and experience and its capacity to "share the fruit of [its] longer experience in the industry." The NOC partner can also support negotiations with and entry into a hard-to-access country. As a senior international downstream manager in the Gulf explained, "We use the label to our advantage. The label implies we give a guarantee of access to our crude." This is particularly useful in Asia, where countries are largely dependent on oil from the Middle East. More concretely, NOCs can help to deal with trade unions, and they understand how to work through a bureaucratic system, such as in China.

There is awareness among the NOCs in the region of the advantages and disadvantages of being associated with their national flag. For instance, governments may support their NOC in commercial negotiations. This advantage is more relevant for winning access to markets than for the exploration and development of an oil field because for the latter, the host government will want the best contractors. Kuwait's strong relationships with Arab states (supported by a tradition of lending money to neighbors) are an advantage for KUFPEC in the Middle East and North Africa, whereas IOCs are viewed with suspicion in these countries. Being the NOC of an Islamic country is also an advantage for companies investing in the Islamic world. As one NOC manager put it, "It's true that being an NOC may be the cherry on top that gets the deals in Iraq or in Yemen. We do have an intimate understanding of Muslim countries." However, outside the Muslim world this may not be an advantage. As noted in chapter 3, affiliation with a Muslim, Arab country was almost a liability for KPI's investing in Europe.

International Strategic Alliances throughout the Supply Chain

As a number of NOCs expand the scope of their activities through the value chain and internationally, they seek to attract foreign investment in new core business and to acquire assets abroad that give them better access to markets and help them to integrate further. There are, in this respect, opportunities for international strategic alliances between IOCs and NOCs, with each party leveraging its assets to access the missing link in its supply chain. Similarly, there is potential for foreign investment to support the greater integration of NOC activities. A Saudi Aramco executive explained that "What would be of interest is a gas value chain. IOCs bring capital and know-how. It would increase opportunities for privatization, which would decrease the funding burden of the government and bring the IOCs in for power generation and the other key needs of the kingdom."

In the natural gas sector, alliances with international private companies present specific advantages, and may be more easily explored because the fixed relations they create lead to interdependence between the upstream and the downstream. The Maghreb–Europe gas pipeline (Algeria–Morocco–Spain–Portugal–France) is an interesting example of South–South and South–North cooperation (Chevalier, 1994: 11).

Several traditional petroleum producers, such as Saudi Arabia, Algeria and Iran, are expanding the focus of their core business; they are turning more to gas in order to support oil exports in generating rent and enabling a greater integration of their activities. There is potential for joint gas development in these producing countries, where the hydrocarbon industry is otherwise closed to foreign investment. Gas is indeed a less politically sensitive resource than oil—note the Saudi Gas Initiative. But when gas products are intended for domestic consumption, subsidies and fixed prices often limit the attractiveness for investors (including NOCs), as is the case with refining. For some time, holders of large gas reserves in the Middle East (except Qatar) saw greater value for gas as a substitute for domestic oil consumption—it would free lucrative petroleum for export—than as a commodity for export. This view is changing. Faced with declining petroleum reserves (in the case of Algeria) and technological advances in gas development, Algeria and Iran want to be the gas hubs of the future, gaining access to markets left open by Qatar and Russia. With regard to the domestic supply of energy too, the scene is changing in the major producers. The Algerian and Iranian governments have put to their parliaments bills aiming to reform the domestic subsidy system, and the Saudi government is expected to review its domestic energy pricing system in order to bring it into conformity with WTO regulations.

Asian companies have been actively pursuing upstream ventures in the Middle East, in strategies often driven by their countries' energy security policies. The Asian importing countries are highly dependent on the Middle East and North Africa for their crude imports: 81 percent for Japan and 39 percent for China (BP, 2005: 18). As a result, Asian governments support their oil companies' efforts to secure new sources of supply in the region. In a parallel move, Middle Eastern producers have tried to gain entry to Asia's downstream markets, primarily in China and India, so as to secure outlets for their crude oil and gas exports and to ensure security of demand.

These forms of interpenetration between Asia's downstream and the Middle East's upstream can be developed further. For example, some Asian national oil companies are exploiting their continued special rights in domestic petroleum activity for commercial advantage. They have retained control of access to markets or control of access to resources. In China, CNOOC has a dominant position in the Chinese market; it is in a position to exchange rights to develop a regasification terminal for access to upstream developments abroad. This interpenetration of upstream and downstream assets makes for a natural fit between importing countries concerned about security of supply and short on reserves and exporting countries preoccupied with security of demand and seeking to expand their markets.

Interpenetration is particularly in the interest of gas exporters, which need secured markets in order to invest in developing their gas resources. Iran is actively pursuing this strategy by offering investments in upstream oil and gas (and petrochemicals) in return for secured gas markets in Asia. In 2005, NIOC announced the signature of a preliminary agreement with Sinopec that will see the Chinese company import LNG over a 30-year period and give it rights to develop the Yadavaran oil field and to import all the field's expected 150,000 b/d output over a 25-year period. Iran has also expressed interest in investing in Chinese LNG receiving terminals. In a parallel development, Japanese companies have pressed on with investment plans in Iran's Azadegan field despite mounting U.S. pressure, and the state-run Chinese oil trading company Zhuhai Zhenrong has come to a preliminary agreement with Iran to import 110 million tonnes of LNG to China over a 25-year period.

On another front, Sonatrach is looking for "new forms of penetration" with IOCs, such as asset swaps. This is an interesting way to develop new skills and to establish a presence in new markets. Sonatrach is discussing an asset swap with Statoil. The companies will exchange stakes of equal value in each other's gas fields. Finding stakes of equivalent value and with similar tax regimes is a major challenge in these ventures.

NOC Buys IOC

NOC ambitions may lead to a new phase of industrial acquisitions in which the NOC purchases smaller private oil companies in a drive to acquire the skills, technology and international exposure that it lacks. This option becomes more viable in the current period of high oil prices thanks to the increased revenue windfall for producers, though assets cannot be bought cheaply until prices fall. An oil super-major would probably be too large for even the largest NOCs to bite off. The capital commitment could not be justified to the state, which invariably has other plans for its revenues. Also, it is likely that, in a reversal of usual roles, the IOC's government would react to its flagship company being bought by an OPEC producer. When KIO, a part of the Kuwait Investment Authority, bought a 23 percent stake in BP in 1987–88, BP successfully appealed to the British government to intervene. BP's board members argued that what was ostensibly a portfolio transaction by KIO was in fact a veiled attempt by KPC to take over BP (Tétreault, 1995: 200–02). More recently, in summer 2005, CNOOC narrowly lost out to American major Chevron Corporation in a bid for the California oil company Unocal Corporation. The Chinese outbid their American competitors but the deal was quashed by political opposition in Washington.[3] There is growing concern in oil-importing countries about this kind of strategic acquisition on the part of a rival oil consumer, and it is likely that perceptions of the OPEC members' ambitions would be equally negative. Political sensitivity in the home country of the IOC could also make the government of the NOC oppose a deal lest it damage diplomatic ties with an ally.

In a less controversial and more viable move, an NOC might acquire an independent. The acquisition of a company with the needed technology and experience might fill the NOC's skills gap. For example, KOC lacks technology and experience with heavy crude, which is set to take a greater share of its petroleum production, and with mature fields, which present ever-greater challenges to offsetting decline. It therefore needs to hone its skills in the production and development of heavy crude, on the one hand, and in enhanced recovery for its existing fields, on the other. It could seek to acquire an independent with extensive experience in heavy crude or a "mid-cap" (medium-sized) international oil company with experience of both heavy crude and mature fields. Saudi Aramco, for its part, needs to secure new refining outlets for its crude, and might benefit from the extensive downstream network of a mid-cap oil company. I sensed a growing interest throughout the Middle East in these strategies.

The challenge posed by such acquisitions would come in the clash of corporate cultures, a problem even in mergers between private sector oil companies. These companies are attractive to the buyer especially because of their lean operations and flexible management processes; and yet, in a merger, these qualities are unlikely to survive in the NOC's operating environment. It is therefore necessary to keep the private company at arm's length and largely

3. Just a few weeks later, another Chinese oil company, the China National Petroleum Corporation, concluded a deal to buy PetroKazakhstan, a Canadian company with oil fields in Kazakhstan, for roughly $4.2 billion.

independent from the NOC while devising mechanisms and processes for exchanging views and developing joint strategies.

Conclusion

Successful new business models of partnerships between NOCs and between NOCs and IOCs will depend on a careful alignment of each party's objectives, needs and assets. This chapter has suggested potential new ways for various types of oil companies to join forces for their mutual benefit. The difficulty lies in finding a venture in which both parties bring complementary assets to the table. Another obstacle is how IOCs and NOCs perceive each other. Neither seems to feel "understood" by the other, and suspicion on the NOC side or frustration on the part of the IOC often mars negotiations. Nor do they fully appreciate how the other has changed. NOCs are unlikely to give turnkey contracts; they want to control and participate in development. This change in NOCs' expectations means that IOCs must either work for the NOC or work with it. By developing new business models, as discussed here, they will become partners.

conclusion

National Oil Companies on the Rise

In the two years since research for this book began in 2003, the oil industry's perspective on national oil companies has changed substantially. Private-sector executives and the financial press now recognize these companies as one of the most dynamic forces shaping the future direction of the industry. In some cases, NOCs are now competing directly with IOCs for projects and investment opportunities overseas, long the preserves of the super-majors. Companies from China, India, Malaysia and Brazil have won concessions to explore for and develop petroleum resources overseas. In today's high oil price environment, they have also been able to leverage their influence to an extent not seen in recent years. As big IOCs scramble to book more reserves in order to convince investors that they have room to grow, NOCs that control access to those reserves have bigger bargaining chips at the negotiating table.

Meanwhile, NOCs are proving themselves able to compete head-on with IOCs in everything from field development to mergers and acquisitions. Saudi Aramco has silenced many skeptics by significantly boosting the kingdom's output without the help of foreign partners. And in spring 2005, CNOOC, the Chinese state offshore oil company, was a serious contender to buy the American major Unocal Corporation, though ultimately it lost out to Chevron Corporation. In the same period, the Iranian National Petrochemical Company was a strong bidder in the competition to acquire a Shell–BASF petrochemical venture, but was unsuccessful partly because of political pressure from Washington on the Shell group not to accept its bid. These ambitious moves show that it is no longer inconceivable that an NOC would swallow up a major oil company.

In the first round of interviews for this study in January 2004, many executives at national oil companies in the Middle East and North Africa had not yet taken stock of this changing environment. At Saudi Aramco, for instance, the term "NOC" was almost a dirty word. Its executives did not even like the company to be included in the same category as other national oil companies. They felt very different from those companies, which they perceived as inefficient, state-dominated and bureaucratic. They would correct me often when I referred to Saudi Aramco as an NOC. Since then, it has taken the lead in several international initiatives by NOCs in an effort to demonstrate its superior management and technical processes, to sell its model to other NOCs and to find high-capacity counterparts with which to engage in joint programs such as lobbying and research. It has, for example, signed up to be the host of the 2006 National Oil Company Forum. Sonatrach was quicker to see the benefits of presenting itself abroad as an NOC. It has been capitalizing on its national status for some years when bidding for assets abroad. Others, such as ADNOC and NIOC, are not very interested yet. KPC is somewhere in between: adventurous and risk-taking in its strategic expansion but lacking the operational autonomy to motivate its personnel and enable it to push forward successfully.

NOCs Are Not IOCs

The five national oil companies have a strong domestic reserves base, unlike most IOCs, which have been struggling recently to find and exploit new deposits of oil and gas as opportunities dwindle. The very large size of the Middle Eastern NOCs' reserves means that they do not necessarily need to develop internationally, as have the private oil majors. They have plenty to do at home, and expansion overseas is much more a luxury than a necessity. Sonatrach may be in a different position: it will need either to acquire new upstream assets abroad or to increase its oil reserves by new exploration and the application of new technology to existing reserves in Algeria—with IOC help as appropriate. NOCs also have an obligation to supply the domestic market with affordable energy and much of the natural gas produced is dedicated to fulfilling this aspect of their mission.

Also unlike the private oil majors, the NOCs do not have extensive downstream networks, nor will they need to develop them. They seek downstream assets essentially to secure outlets for their crude and to be closer to markets, and for these reasons they do not need to expand their integration to the levels reached by the private companies. However, when an NOC such as the Kuwait Petroleum Corporation derives its main source of funds from downstream activities, its level of integration will be closer to that of IOCs.

Financially, national oil companies remain different creatures than the private oil companies. Their finances are not independent of government. This is especially true of NIOC because the state regularly draws on its funds, which limits its capacity to carry out investment programs. NOCs are also subject to government oversight of their spending. Their "owner" the state is not like a private owner: neither the NOC nor the state has an immediate alternative to the other.

NOCs are different from IOCs too in their long-standing engagement in the promotion of social welfare, for example through educational programs, infrastructure development, local procurement and new private-sector business development. Over time, they are disengaging from this type of expenditure, but their historical concern with the economic impact of their investments on society has left an imprint on their thinking about their obligation to society.

Fundamentally, they are not like international oil companies because they have a monopoly or near-monopoly over their countries' resources and do not have a majority of private shareholders. A number of state companies have experimented with a degree of private ownership, offering shares to the public while allowing firm government control to be maintained. Models such as Statoil, Petrobras, Pemex and Russia's gas giant Gazprom, which is traded on the stock exchange but is now under strong control by the Kremlin, have grown into new entities that have characteristics of both an NOC and an IOC.

State ownership does not preclude NOCs' adherence to the more transparent and rigorous accounting standards of private companies. Sonatrach is a case in point. Its accounting adheres to generally accepted international standards and its annual report makes disclosures of financial accounts. Also, some national oil companies publish data on their reserves classification. This is the practice of Pemex in Mexico, which is required to conform to U.S. Securities and Exchange Commission regulations on disclosure of geological data because its securities are traded in the U.S. markets.

However, a distinguishing characteristic remains: national oil companies are instruments of the state. Their operations and strategy are restricted by government directives. For instance, they are required, for the most part, to use their international refining assets as outlets for national crude, even when this is uncommercial. They must also respect employment directives that raise operational costs. IOCs must often respect similar labor directives, but they will usually require other gains in order to offset those directives' impact on their return on investment. In other words, NOCs do not always operate on the basis of a commercial rationale. They may serve the state's strategic interests and its social welfare objectives as well as the more common objective of the oil and gas business of generating profits.

NOCs Are Unlikely to Become IOCs

Outside the producing countries, many people think NOCs should be privatized and that their governments and societies would benefit most from such a step. This view is heard in the halls of the World Bank and the International Monetary Fund and is expressed by a number of IOC executives and government officials in the importing countries. But it is not the predominant view in the producing countries. NOCs serve state interests more directly than do private companies because they are instruments of the state. Moreover, there are strong and enduring popular expectations regarding the national status of these companies. This was evident in public responses to the hydrocarbon sector's reform initiated in Algeria: civil society groups strongly voiced their opposition to any possibility of privatizing Sonatrach. It is also apparent in the way in which the Kuwaiti and Iranian media comment on the national oil and gas industry's performance. Their criticism is far from muted, but it falls short of calls for privatization.

Another reason why NOCs in the Middle East and North Africa are not about to become IOCs is that they have a developed a specific corporate culture, which is imbued with their identity and their national culture. Though they talk of their ambition to emancipate themselves from the confines of the NOC model and succeed like an IOC, they derive a great deal of pride from their domestic status. This was evident in my interviews with young professionals. They are very attached to their company's status and role. They are proud to work for the national flagship company.

Becoming a Top-Notch NOC

However, there is no reason why their national status should prevent any of these companies from being highly competent and efficient. The five companies under study can strive to excel within the national oil company model.

This involves developing NOC and company-specific strengths and addressing weaknesses. Their strengths come from their relationship with, and support from, their government and society. The more harmonious these relationships are, the stronger the national oil company will be. Furthermore, the state's assets are strengths of the national oil company because it usually enjoys exclusive or preferential rights on home territory. These assets include, besides natural resources, the geographic location of the country and the government's network of alliances and relations with other countries, potential hosts of the NOC's overseas development.

In most cases, their weaknesses pertain to poor processes, both internally and in their relationship with government. Consultation is a traditional decisionmaking process in the Middle East. But internally, many companies have centralized, top-down and rigid decisionmaking processes, which hinder their responsiveness to new trends. Management could introduce more consultative, consensus-building processes instead, as ADNOC has done.

Another weakness of regional national oil companies concerns the burden of national mission expenditure. However, most governments will willingly take on the companies' social programs. This transfer of national mission to the state has already been accomplished in Iran. Ideally, NIOC will retain the capacity to develop programs that benefit its own strategic objectives—as Saudi Aramco and Sonatrach have demonstrated with investments in training to build national capacity, which increases the pool of skilled labor for the company to recruit from. They and NIOC work with local content policies that increase business for the national private sector, and this ultimately reduces the dependence of the non-hydrocarbon economy on hydrocarbon revenues. As these NOCs build up the national capacity needed to support the oil and gas business, they contribute to their countries' economic welfare and thereby increase their national prestige. It is therefore possible to turn this weakness into a strength.

A key factor for success within the NOC model will be to improve accounting processes that waste government revenues or that leave the national oil company unaccountable. At least internally, clear financial accounting is needed in order to make costs and profits explicit. More generally, a clear accounting of activities will allow benchmarking between NOCs. The benchmarking of cost efficiency or HSE standards would be a valuable way for governments to assess the performance of the national industry. It would also be useful for the national oil company, which often operates in a monopoly (or close to one) on the national territory and lacks the pressure of competition to drive its performance. In fact, national oil companies are keen to compare their performance to that of other companies. So far, it has been almost impossible to conduct any substantial benchmarking exercises because the NOCs concerned are usually prevented by government from disclosing financial and operational data.

Governments will need to collaborate in efforts aimed at resolving efficiency issues, increasing the operational autonomy of national oil companies and improving the governance processes of the oil and gas sector. To ensure government cooperation, companies must return to the state its policy prerogatives. Moreover, their leaderships must be clear about the demarcation of roles and responsibilities between the NOC and relevant state institutions.

It is also crucial for the state and society to recognize that the national oil company cannot be expected to solve the country's economic problems on its own. It does not have the capacity or the responsibility for this. The state must make sustained efforts to diversify the economy and avoid a situation in which, 15 to 20 years from now, it is so dependent on the national oil company for revenues that it smothers it. An NOC free to operate commercially, within the boundaries delineated by state oil policy, will establish the reign of the oil titans for years to come.

economic background

The Challenges Faced by Petroleum-Dependent Economies

Special Contribution by John V. Mitchell

Unlike international oil companies, which can choose where to invest or disinvest, national oil companies are bound to their national operating environment. In their turn, the economies of major petroleum exporting countries are bound to their NOCs. There are two structural faults in this interdependence: petroleum production will inevitably flatten and then decline, and petroleum revenues are unstable. To manage these two problems, countries require a strategy for the development of the non-hydrocarbon economy (to generate tax revenue, employment and exports) and a buffer between hydrocarbon revenues and the continuing and more stable revenue needs of the government. If these requirements are not met, the oil revenue delivered by the NOCs will not be sufficient to counter the problems discussed in chapter 5, such as increased unemployment and a decline in living standards, which could ultimately erode the sociopolitical contract and cause dissent within society.

Economic Reform

During the mid- to late 1990s, a combination of relatively static oil revenues and rising demand for jobs led to chronic budgetary deficits in the major oil exporting countries. Governments were forced to embark on economic reform programs. The first objective was the classic one of balancing their budgets. Most governments required IMF assistance in order to restructure their external debt, and the IMF promoted the economically orthodox but controversial "Washington Consensus" policies for the oil exporters, as for other developing countries and the emerging market economies of the former Soviet Union. This consensus, though never complete, especially regarding timing and sequencing, involved several elements: fiscal discipline; tax reforms; directing public expenditure toward infrastructure and public services rather than industrial activity; privatization of state industries; banking reform, including liberalizing interest rates; deregulation of markets; competitive exchange rates; trade liberalization and opening up to foreign direct investment; and securing property rights.[1] Most of these elements are reflected in the reform programs of the oil exporting countries. Government spending has been brought into line with expected revenue, and business conditions for the private sector have improved. None of the five countries in this study intend to privatize its national oil company, however.

For the petroleum exporters of the Middle East and North Africa, there is an additional driver of reform: the high growth rate of the population, and, with it, the even higher growth rate of unemployment among young people of working age. The reform programs have reduced the growth of employment in the public sector and state enterprises. The hydrocarbon sector will never be a large employer. Jobs in the medium term, as well as tax revenue and export earnings in the longer term, must therefore come from the growth of the private non-hydrocarbon sector, though in Kuwait and the UAE, where nationals form only 35 percent and 25 percent of their respective populations, governments have an option in the longer term to reduce, halt or reverse the growth of the expatriate labor force.

All the five governments recognize the importance of diversifying their economies and reducing their dependence on petroleum revenues, but the priority given to providing employment opportunities for nationals often overshadows the responsibility to design policies for developing new sources of revenue and foreign exchange.

The critical question for the hydrocarbon sector is how far it can deliver growing support to the dependent economy in the medium term. Except in Saudi Arabia, it is difficult to increase oil exports (and, in some countries, gas exports) and support growth in petroleum revenue without expanding capacity. As was made clear in chapter 8, all the exporters have plans for such expansion. With simplifying assumptions, it is possible to make a rough guess

1. For a review by the originator of the term "Washington Consensus," see John Williamson, "What Should the World Bank Think of the Washington Consensus?" *World Bank Observer*: 15 (2), August 2000.

at how long the dependence can be sustained by expanding oil and gas revenues under different scenarios. Though the answers differ from country to country (because of their different resource endowments and non-petroleum potential), the general answer is that the first crunch point will come when oil and gas production reaches a plateau. Then, the hydrocarbon sector's ability to support further growth in the non-petroleum economy will also plateau. For the exporting countries, this means within the next 10 to 20 years. Prior to that, very large and escalating adjustments will be needed in order to reduce the non-petroleum economy's dependence on oil unless prices remain at their present (late 2005) high levels through the medium term and rise even further in the longer term when the second crunch point, declining production, occurs.

This chapter outlines the macroeconomic predicament and future challenges facing each of the five countries. It then considers the hydrocarbon sector's contribution to their economies, with reference to the results of simulations on meeting the non-hydrocarbon fiscal deficit (NHFD) and balancing the current account. Finally, the demands that these challenges will make on the hydrocarbon industry are discussed in the context of the broader need for strong economic policy initiatives. The scenarios are based on conservative price assumptions that follow from a long-term view of average oil prices. The recent (2005) increases in international oil prices have already created a financial and foreign exchange buffer for all these countries. If prices were to settle at a level above those of our scenarios and if spending in the non-oil economy were unchanged, the effect would be to defer the crunch points. But in their 2004–05 budget assumptions, most countries already plan to increase their expenditure, and the prospect of longer-term high prices will inevitably lead to even higher deficits of tax and foreign exchange in the non-hydrocarbon sector in the future than are projected here. These countries' dependence on hydrocarbon revenues and export earnings will increase. The inevitable plateau of production will create the crunch points we describe. Thus, even with higher revenues (and export volumes restricted by the effect of higher prices on export demand), their timing may not change very much.

Economic Background

The Kingdom of Saudi Arabia

During the 1980s, the Saudi economy withstood the oil price explosion of 1979–80, the drop of oil demand leading to the collapse of oil prices in 1986, the slow reconstruction of the kingdom's share of the world market and the first Gulf war. Foreign balances, which had been built up in the early 1980s, were run down as revenues fell, and were finally diminished by Saudi Arabia's contributions to the cost of the war.

In the 1990s, the government introduced tighter controls on its spending and began to eliminate subsidies and distortions in the pricing of oil and utilities to domestic consumers. However, oil revenues were roughly static, with offsetting price and volume changes. Foreign balances continued to be run down until 2000.[2] Though the non-oil budget deficit fluctuated, it remained more or less constant at just below 200 billion riyals ($53.4 billion) in real terms. At this time, the non-oil economy grew at about three percent, so that the ratio of this deficit to the non-oil economy fell. However, in 2002 and 2003, it was still 41 percent, the highest of any of the countries studied except Kuwait.

Meanwhile, the rate of population growth, though lower than in the 1990s, has been estimated at 2.7 percent for 1990–2003 (World Bank, 2005a: 50, table 2). Government spending in total has continued to grow at 3.7 percent, and capital expenditure at 2.7 percent. In 1992–2002, personal consumption was squeezed, falling on average by 1.5 percent every year in current terms. Capital expenditure has remained around 20 percent of GDP. This is not high compared to other developing countries of similar income per capita but without oil.

In 1994, the government began a program to restructure the economy and increase the scope of the private economic sector. As part of the Foreign Investment Act of 2000, it established the Saudi Arabian General Investment Authority to design and carry out the privatization program. Two years later, 20 sectors had been earmarked for total or partial privatization, including Saudi Arabian Basic Industries Corporation (SABIC), the Saudi Arabian Mining Company (Ma'aden) and local oil refineries that currently have foreign private shareholdings. The Foreign Investment Act authorized 100 percent foreign ownership of enterprises in all but 16 sectors, which included oil exploration, drilling and production, and petroleum distribution and retail services of all kinds.[3] The 2003 Gas Supplies Investment Law and the Natural

2. This was mainly due to the central government's offering attractive interest rates to local banks, commercial enterprises and individuals. These parties withdrew capital from abroad so as to gain higher (and tax-free) returns.

3. Prior to this law, foreign companies were limited to a 49 percent share of joint ventures with Saudi domestic partners.

Gas Investment Taxation Law provides for foreign investment in the gas sector.[4] Foreign investors will pay a tax of 35 percent on profits and will be eligible for loans from the Saudi Industrial Loan Fund. New stock market regulation and legislation is also facilitating the development of the equity market in the kingdom.

Saudi Arabia already grants tariff preferences to other GCC countries, and the GCC states have begun to establish a full customs union. In November 2005, the kingdom received formal approval for WTO accession, which will create further obligations to separate state- and private-sector activities, increase domestic fuel prices and reduce discrimination against foreign investors and customers. Bilateral trade agreements have been concluded with the EU and many other countries.

In parallel with these structural economic reforms, the kingdom has set out ambitious targets for improving educational standards, and therefore the employability, of the younger generation. The labor market remains distorted, however, by the high salaries paid in the public sector (which employs 74 percent of Saudi nationals in the labor force) and the low pay of most jobs performed by the 3.7 million expatriate workers. In 1999, wages for Saudis were around three times higher than wages for non-Saudis. In production, they were 2.9 times higher, but wages were similar in managerial positions (Diwan and Girgis, 2002). The rights and conditions of expatriate workers are being improved, in parallel with an aggressive policy of Saudization in which the hydrocarbon sector plays an exemplary role (Pakkiasamy, 2004).

After a decade of running deficits, the high oil prices of 2003–05 have allowed the Saudi government unexpected budget flexibility. In 2004, there was an increase of $10 billion (15 percent) in expenditure, mainly on social infrastructure and the salaries of the security forces. Domestic debt was reduced by $11 billion. The central bank's foreign assets increased by $28 billion in 2004; and by November 2004, it provided 28 months of import cover. Though the domestic debt was reduced, some windfalls are likely to have been exported by the banking system and other domestic lenders in order to acquire foreign assets, as happened in 2002–03. The fiscal stimulus, together with the effect of economic reforms, supported a growth rate of 5.7 percent in the non-oil private sector GDP in 2004 (Saudi American Bank, 2005: 14).

4. The gas sector remains under the direction of the Supreme Petroleum Council, with Saudi Aramco as the executive agency.

The Islamic Republic of Iran

Iran underwent much economic turbulence in the 1980s and 1990s. Oil production fell from 5.7 million b/d before the 1979 revolution to 1.3 million b/d in 1981, though the effect on revenue was offset by the price increases of the second oil shock. The rebuilding of production was limited in the 1980s owing to the war with Iraq and also OPEC quota cuts, which aimed at limiting competition as oil demand collapsed after the price shock. During the 1990s, production increased steadily, reaching 3.9 million b/d by 1998. The oil industry's difficult years were reflected by the contraction of the Iranian economy in the 1980s. But even though the Iranian economy shrank during the period 1977–88, over the longer period, from 1960 to 2002, the annual rate of growth was 4.6 percent. This was higher than the average of other Middle Eastern and North African countries and of every individual country except Qatar and the UAE (IMF, 2004b: table 1-1). If it were not for structural inefficiencies in the economy, higher growth rates could have been expected given the record of investment and the increasing number (and educational standard) of the labor force (IMF, 2004b: table 1-4).

Following the 1979 revolution, the economy was in effect under the control of the central government. But a large section of it came into the hands of non-state (religious) institutions, which enjoyed a combination of monopoly in certain sectors and independence from the government. Priority was given to social programs, especially education and health. Since 1978, the average level of schooling of the working population has tripled (though it is still only five years); life expectancy has increased and adult illiteracy has dropped from 40 percent to 20 percent. Investment has recovered in recent years to its 1960–2002 average of about 30 percent of GDP. This is similar to that of the rapidly growing East Asian countries, but the productivity of the investment appears to be lower. Economic development has been distorted by central controls, high tariffs (still at around 30 percent), subsidies on energy consumption and import subsidies, whose cost was met in part by the multiple exchange rate regime.

Financing the expenditure program by short-term borrowing led to a financial crisis in the mid-1990s. With the help of the IMF, debt was restructured, the external and internal deficits were reduced and inflation was brought down to the current annual rate of about 15–16 percent.

Since 1990, the Iranian economy has been directed as far as possible by a series of five-year plans (the third plan ended in March 2005) in which the state's management role has been reduced in favor of market mechanisms.

During the third plan, real per capita GDP and real per capita private consumption increased. Iran has generated small surpluses on its current account and has a very low level of external debt. Since 2000, it has directed fiscal surplus from oil revenues above $15/b into a stabilization fund that may be used, with express parliamentary approval, for investment in the nonhydrocarbon sector. Exchange rates were unified in March 2002, and there has been an extensive reduction of barriers to imports and exports. Import tariffs have been reduced to 45 percent and corporate taxes to 25 percent. In addition, domestic taxes have been simplified and their collection has been improved. A value added tax has been proposed to parliament, to take effect from 2006–07, and a gradual phasing out of implicit energy subsidies is planned.[5]

The primary role in generating investment and employment has shifted from the government to the private sector, helped by a privatization program.[6] Furthermore, gross foreign direct investment, mainly through buybacks, reached about $3 billion in 2003–04 (IMF, 2004a: table 4). In its 2004 review, the IMF drew attention to the need for Iran's stricter adherence to the stabilization fund's objective of countering oil price volatility (rather than increasing investment during periods of high oil prices). It also noted the need for a specific plan to reduce energy subsidies and for further reform of the banking system, to enable it to function more effectively in financing private-sector development.

In the economic sphere, the principal challenges for Iran in the next decade are:

—To maintain and extend the reform process to support a higher growth rate by eliminating identified slacks and distortions in the economy (mainly energy subsidies);

—To continue privatization and decentralization of public-sector economic activities;[7]

—To reduce the growth of domestic oil consumption;

5. Progress on these reforms appears to be slow. In January 2005, the Majlis decided to freeze domestic prices for gasoline and other fuels at 2003 levels, with money from the Oil Stabilization Fund being allocated to cover the rising cost of imports (EIA, *Country Analysis Brief—Iran*, last updated March 2005).

6. A general law on foreign capital, which applies to the petroleum sector with special provision, was introduced in 2002. This updated the 1956 Law on Attraction and Protection of Foreign Investment by adding to the areas in which foreign investors can enjoy the same protection and privileges as national investors and includes special provisions for petroleum investment.

7. See World Bank, 2003b: 149–69, for a thorough discussion.

—To reduce unemployment through higher growth and labor market reforms. The fourth five-year plan, now in preparation, is likely to aim at a GDP growth rate of 8 percent in order to keep unemployment below 10 percent (in 2002, it was reported to be at 11 percent);[8] and

—To shift the economy from its dependence on oil and gas exports, in anticipation of the eventual plateau and later decline in oil and gas production.

People's Democratic Republic of Algeria

As discussed in chapters 1 and 2, hydrocarbon development has contributed politically and financially to a difficult political history in Algeria. Following nationalization of foreign interests in the oil and gas sector in 1971, the FLN one-party government in Algeria operated a command economy without restraint until 1986, when political frustration triggered a period of civil war and a contested transition to a multiparty government. This coincided with the collapse of the world oil price and a fiscal and financial crisis resulting from the accumulated imbalances in the rest of the Algerian economy. Per capita GDP fell, and only in 2003 did it return to the level of 1990.

Some economic restructuring took place in 1989 and 1991, including a fundamental relaxation of restrictions on foreign investment in the hydrocarbon sector. A comprehensive macroeconomic stabilization program began in 1994. This reversed the unsustainable accumulation of budget deficits and external debt and began a process of economic reform and liberalization. The negative trend in productivity in the non-hydrocarbon economy was reversed, and there was a reduction in the proportion of the population in poverty. Inflation was reduced and has been stabilized. However, job cuts in the public sector have added to unemployment since 1996. This has remained high (27 percent of the labor force, and much higher still among the young) despite the resumption of economic growth and the slowing down of the annual rate of population increase, from 2.4 percent to a predicted rate of 1.5 percent. National, provincial and presidential elections in 2002–04 reflected political reforms, which have strengthened the government's legitimacy. The EU–Algeria Association Agreement, signed in December 2001, further underpins the positive direction of both political and economic reform, committing Algeria to reducing trade barriers and introducing modern legislation on competition and the protection of intellectual property.

8. This information is from an interview with Mohammad Kordbache, Director-General, Office of Management and Planning, Netiran, 22 November 2003.

The World Bank reports that the volatility of hydrocarbon revenues for Algeria has been significantly worse than for most other oil exporting countries.[9] The effect of this volatility has been magnified by a public fiscal policy that protects consumption during periods of low revenues, at the expense of capital expenditure. Fluctuating capital expenditure may partly explain why the country has grown more slowly than other oil producing countries in the Middle East and North Africa since 1989. It should also be borne in mind that, until very recently, Algeria has had less scope than other exporters to compensate for lower prices by increasing volume.

To counter this volatility, Algeria set up a stabilization fund (Fonds de régulation des recettes) in 2000 into which the government deposits any revenues from an excess of hydrocarbon prices over the budget forecast. These funds are reserved for budget stabilization and investment. However, payments into and from the fund are not published, and the budget's price assumptions have been revised upward during recent price increases. As a result, it is uncertain how far this mechanism acts a) as a constraint on spending, b) to sterilize the inflationary effect of surpluses by investing funds abroad and c) to build up assets against future oil depletion. Capital investment in Algeria itself, currently at around 35 percent of GDP, is similar to the average for the Middle East and North Africa and for countries of Algeria's income level. However, it is lower than that in Middle Eastern petroleum exporting countries.

Although the reforms of 1996 and their follow-through have increased fiscal and macroeconomic stability, the legacy of "command and control" government continues to cause serious economic problems. Future challenges to Algeria will be:

—To cut unemployment. The IMF estimates that an annual rate of growth of six percent (in the whole economy) would be necessary to halve the unemployment rate over a ten-year period;

—To address the weakness of the predominantly state-run banking system;[10]

—To increase competitiveness of manufacturing industries by reducing protective tariffs (the EU–Algeria Association Agreement of 2001 requires the gradual reduction of tariffs over a 12-year period);

—To increase transparency in government financial and administrative operations;

9. Recent developments in Algeria and its medium-term prospects have been assessed by the World Bank (World Bank: 2003a).

10. State-owned banks in Algeria carry a burden of bad loans (and implicitly weak solvency ratios). Many of their debtors are state-owned enterprises.

—To enable the "old" non-hydrocarbon economy to survive in the transition to the market economy. This will require reducing its fiscal deficits and providing foreign exchange to pay for a greater share of its imports. It will involve, among other steps, speeding up privatization and attracting foreign direct investment.

The government has devised policies aimed in this direction, but bureaucratic processes, distortions of access to capital and some residual political violence are deterrents to investment.

The State of Kuwait

The economic structure of Kuwait, apart from its dependence on oil, is defined by its demographics. In 2004, the public administration employed 79 percent of all employed Kuwaitis. Participation in the labor force is at about 30 percent for Kuwaitis and over 70 percent for expatriates, with 95 percent of the Kuwaiti labor force and 99 percent of the expatriate labor force in employment (IMF, 2005c: tables 10, 11).

The expansion of the labor force, estimated at around six percent per year, reflects a high level of immigration (the expatriate population increased by 11 percent in 2004, according to the National Bank of Kuwait) as well as a trend of about a three percent per annum expansion of the Kuwaiti population. Unlike in the UAE, the expatriate labor force does not form the basis of significant export- and re-export-oriented commercial activity. This is because non-oil exports make up only some five percent of the non-oil GDP. The expatriate population fluctuates with the state of the economy. There is no necessity, at present, to assume that it will continue to expand at recent rates.

Like other GCC countries, Kuwait has recently begun a process of structural reform intended to reduce the role of government in the economy and to liberalize labor, capital and goods markets and the general conditions for private-sector business. Since October 2003, social security benefits for Kuwaitis working in the private sector have begun to be brought into line with those of Kuwaitis in the public sector; this should make employment in the private sector more attractive and help private companies to reach their "Kuwaitization" quotas. The 2001 Foreign Direct Investment Law allows 100 percent foreign ownership of many types of business and sets out how the rate of corporation tax (applied only to foreign companies) will be lowered. There has also been a range of measures to improve the transparency and efficiency of the banking and capital markets. A privatization law has been proposed that would allow the divestment of major public-sector utilities. Many of the government's policy intentions are still to be implemented, however. In the hydrocarbon sector, Project Kuwait, which seeks to engage foreign companies in a limited area of exploration and development, has been under discussion for over 10 years, and is currently again before parliament.

It is difficult to project the economic trends and structure of Kuwait into the distant future without some vision of what population structure—and implicitly what kind of society—is the objective. Today, two-thirds of the population (and 80 percent of the labor force) are not Kuwaitis. With the economic imbalances in the labor market that go with this population structure, the non-oil economy of Kuwait is likely to remain dependent on oil revenues, both for government revenue and foreign exchange, to a higher degree than will be the case with other major oil exporting countries. This dependence might be reduced if the growth of the expatriate population were limited or reversed, but this would require replacing their services with high-cost Kuwaitis and therefore a shift in the non-hydrocarbon economy toward higher value added business.

The size of Kuwait's oil and financial reserves will allow its dependence on crude to continue for another generation or more, depending on the price of oil, but a reduction in this dependence is not in sight. Despite the Fund for Future Generations, which takes 10 percent of Kuwait's oil revenue, the savings rate is not high (about 20 percent), and domestic investment absorbs only about half these savings. There is a trend toward liberalization and opening business sectors to foreign companies, which can bring expertise and possibly markets, as in the case of the Equate petrochemicals project. However, the manufacturing sector remains small, employing only six percent of the total labor force in 2002 (and only 2.7 percent of the Kuwaiti labor force). Unlike the UAE, Kuwait does not appear to be setting up a manufacturing base aimed at using cheap migrant labor for assembly and processing for neighboring, less open markets.

In addition to finding a similarly appropriate basis for its non-oil economy, the key challenge to Kuwait is to develop a vision of its future demography. Only then can the government develop strategies to accommodate society's future economic needs.

United Arab Emirates

The Emirates are dependent on oil (and continuing oil production relies on the reinjection of gas), but several features differentiate them from most other major hydrocarbon exporting countries:

—A very high endowment of both oil and gas resources per capita (varying widely between the constituent emirates. Abu Dhabi, the most populous state, also has the highest reserves per capita, more than twice that of Saudi Arabia);

—A majority non-Emirati population, which is the basis of a rapidly growing non-hydrocarbon sector aimed at re-export, assembly and services for neighboring countries where conditions for private-sector business are less liberal;

—A federal political structure with decentralization of authority; and

—Continuous participation of major international oil companies in the oil and gas sector, so that ADNOC, though it has a controlling role, does not have a monopoly.

Compared to other oil producers in the Middle East, the UAE has had less need to dismantle or rationalize vast state enterprises or to reform domestic energy prices. Economic reforms in the 1990s were directed at making an already pluralist economy work more efficiently. There are many options for development both within and outside the hydrocarbon sector. The demands on ADNOC will depend on what choices are made among these options regarding the population and the non-oil economy.

The Emirates are also developing non-hydrocarbon economies for which a sustainable future can be imagined. As the GCC customs union takes effect, the Emirates' industries are likely to be strong competitors in the regional market. However, fixed investment in the non-hydrocarbon sector is not high—30 percent of the non-oil GDP, and its expansion, has been based on the increasing supply of cheap migrant labor and cheap energy.

Eventually, the non-hydrocarbon sector's degree of fiscal dependence on the hydrocarbon sector will be incompatible with the rate of population expansion. A slowdown in the rate of the non-hydrocarbon economy's growth (and therefore of employment) could be accommodated by slowing or even reversing the flow of migrant workers. The population could be adjusted to fit the economic resources available—this would alter the balance of cheap versus expensive labor—and the slowdown would affect sectors such as construction and service industries in which nationals have controlling interests.

This dependence could be alleviated by persevering with existing policies designed to improve the efficiency of the public sector and the ease of doing business in the private sector and to raise the technical and managerial capabilities of the long-term population. However, there are challenges: re-export and free-zone business activities will face competition as their regional markets become more open and their own local industries become more efficient. The Emirates have yet to develop high-value-added exports. The continuing success of their non-oil businesses will need "more that is different" rather than "more of the same." The pressure is somewhat different on Abu Dhabi, the richest in petroleum resources, and the other emirates, for which the non-hydrocarbon sector is already more important.

The Hydrocarbon Sector's Contribution

The hydrocarbon sector dominates the economies of the five countries, as the simple statistics in table 1 show.

As with all countries, what happens in the long term depends on achievements in the governments' core activities of law, fiscal discipline and institution building and on the successful continuation of market reforms. In the medium term, the hydrocarbon sector in each country has a clear and positive role to play in expanding oil and gas production and exports and thus in increasing government revenues and the country's capacity to import goods and services. Led or monopolized by the NOC, the hydrocarbon sector also contributes to the rest of the economy in other ways, such as through dynamic linkages and cheap energy. However, the nature and impact of these contributions will vary from country to country, as described below.

Dynamic Linkages

All five NOCs contribute dynamically to their economies through local content and national employment policies. They may give the private sector incentives to develop, in the form of a margin of preference allowed in evaluating bids (such as in Saudi Arabia) or a quota. For example, NIOC's buyback agreements with foreign partners require 51 percent of expenditure under the agreement to be placed with Iranian contractors, manufacturers and consulting firms (these may include Iranian companies with some foreign participation). ADNOC and its consortia—that is, all the oil producing companies—are required to obtain 51 percent of their goods and services from the Emirates. NOCs also provide training and education in technology for hydrocarbon-sector employees and, in the case of Saudi Arabia and Iran, for the local contractors who supply the industry.

These contributions are qualitatively and strategically important; but in macroeconomic terms, they are small compared to that of the large and

Table 1. The Hydrocarbon Sector's Contribution to the Five NOC States, 2003
Percent

	Saudi Arabia	*Iran*[a]	*Algeria*	*Kuwait*	*UAE*
GDP	42	22	36	51	38
Exports of goods	88	80	98	92	80[b]
Government revenue	78	60	69	77	81

Sources: Saudi Arabia: Saudi Arabian Monetary Agency (SAMA), 2003b, Tables 9.1, 8.2, 5.1; Iran: IMF, 2004a, Tables 3, 4, 7; Algeria: IMF, 2005b, Tables 2, 15, 30; Kuwait: Central Bank of Kuwait, 2003, Table 80, Central Bank of Kuwait, 2004, Tables 46, 35; UAE: IMF, 2004e, Tables 5, 19, 37.

a. Iranian solar year 1382 (21 March 2003–19 March 2004).

b. This figure is for net exports. For gross exports from the UAE, including re-exports and exports from the Free Zone, the figure is 42 percent.

growing labor forces of most of these countries. The hydrocarbon sector directly employs only 1 to 2 percent (less in Iran, because of its larger active population) of the total labor force in most countries.[11]

The problem for the oil industry comes, as pointed out in chapter 6, when pressure on the NOC to employ ever greater numbers may constrain its commercial potential. KPC has faced this pressure in the past, and continues to deal with its reverberations. There is likely to be more pressure on Saudi Aramco, despite Saudi Arabia's comparative advantage in terms of production and export potential.

Cheap Energy

The NOCs in these countries are monopoly suppliers of "cheap energy" to the domestic market, through oil and gas prices controlled (sometimes indirectly) by government. Iran and Algeria achieve this through subsidies provided by the NOC, the government, or both. In Saudi Arabia, Kuwait and Abu Dhabi, fuel is simply cheap.[12] Feedstock for the power and petrochemical sectors is also obtained for less than market value, but such transactions are less transparent. Underpriced supplies are an "invisible transfer" from the oil sector to the non-hydrocarbon sector. The efficient alternative would

11. A rough guide to oil sector employment as a percentage of the total labor force is: Algeria 1.7 percent (direct contribution); Saudi Arabia, 1.4 percent direct and 5 percent indirect; ADNOC 1.2 percent (oil sector, excluding natural gas and processing); Kuwait, 1 percent (KPC and subsidiaries); Iran, less than 1 percent direct. Sources: ILO LABORSTA data (*http://laborsta.ilo.org/*) and *Saudi Aramco Facts and Figures 2002*).

12. Electricity is subsidized in Iran and Kuwait. In Kuwait, it is subsidized through the budget. In Iran, the way oil and gas are priced for electricity generation implicitly carries the electricity subsidies.

entail bringing domestic oil prices in line with export prices, with adjustments for differential costs. The higher prices would yield the government revenue on the domestic markets at near the rate received on exports.

Though oil and gas subsidies contribute to the non-petroleum economy, they also distort it and hinder its growth. The removal or reduction of subsidies would benefit the economy in the long term by increasing efficiency and generating additional revenues that the government could invest in the non-hydrocarbon sector. For example, revenues forgone through low domestic fuel prices in Iran and Saudi Arabia were probably equivalent to 9–14 percent of their non-petroleum GDP. In Algeria, natural gas is the dominant fuel and the main carrier of subsidies. If the average government tax take of 70 percent of the gas export price were applied to domestic sales, an extra 60 billion dinars ($800 million) could be generated.

In the short term, however, removing subsidies could damage parts of the non-oil economy. For example, fuel price increases would affect the international competitiveness of Algeria's and Iran's energy-intensive industries. These industries already enjoy the protection of relatively high tariffs, which impose costs on the domestic market; and in Algeria, they contribute little to exports. Because gas subsidies give the Algerian industry an artificial cost advantage, they will be phased out over 12 years under the EU–Algeria Association Agreement. In the household sector, price increases would have a serious impact on low-income groups, for whom energy consumption is a high proportion of the domestic budget.

Government Hydrocarbon Revenue and the Non-Hydrocarbon Economy

In all five countries, government expenditure exceeds revenue outside the hydrocarbon sector by a large margin. The proportion of the non-hydrocarbon fiscal deficit (NHFD) to the non-hydrocarbon GDP is a measure of the adjustment that will be necessary to protect the overall fiscal balance from the flattening and eventual decline of oil revenues.

In table 2, the ratio of the NHFD to the non-hydrocarbon GDP is presented, together with the general fiscal balance (what remains after petroleum revenues have paid the fiscal deficit of the non-hydrocarbon sector). The picture in 2002–03, when oil prices were close to their post-1986 average, was of governments broadly controlling their expenditure in order to achieve a rough fiscal balance. This kind of balance is necessarily more vulnerable to falls in petroleum revenues, but the picture is blurred by special structural factors, which make comparisons inexact.

Table 2. Oil in the Fiscal Balance of the Five NOC States

Country	*Year*	*Non-hydrocarbon fiscal deficits as a percentage of non-petroleum GDP*	*Non-hydrocarbon GDP ($bn)*	*Non-hydrocarbon fiscal deficit ($bn)*	*Fiscal balance ($bn)*
Saudi Arabia	2002	41	121	50	–5
	2003	41	127	52	10
Iran[a]	2002	19	112	22	–3
	2003	19	120	23	0
Algeria	2002	32	38	12	1
	2003	35	42	15	3
Kuwait[b]	2002	42 (78)	20	9 (16)	10 (2.2)
	2003	49 (83)	23	11 (19)	10 (2.5)
UAE	2002	35	47	17	8
	2003	36	50	18	11

Sources: Saudi Arabia: SAMA, 2003b, Tables 9.1, 5.1; Iran: IMF, 2004a, Tables 3, 7; Algeria: IMF, 2005b, Tables 2, 15; Kuwait: Central Bank of Kuwait, 2003, Table 80, Central Bank of Kuwait, 2004, Table 35; UAE: IMF, 2004e, Tables 5,19.

a. Iranian solar years, 1381 (21 March 2002–20 March 2003), 1382 (21 March 2003–19 March 2004).

b. Figures in brackets exclude income from the Reserve Fund for Future Generations and the General Reserve Fund.

In Saudi Arabia, there are significant extra-budgetary flows of revenue and expenditure, including Saudi Aramco's operating and capital budgets, the financing of cheap oil products in the domestic markets and some defense expenditure.

In Iran, NIOC's operations are consolidated into government revenue, but the cost of low petroleum prices is shown neither in the government budget nor in the GDP figures (this treatment of those figures is due to change). If oil and gas subsidies, estimated by the IMF at 117 trillion rials ($14.2 billion), had passed through the government budget and been included in the figure for non-petroleum GDP expenditure at market prices, the figure for the contribution of oil and gas to government expenditure in 2003–04 would have been higher. As indicated in chapter 7, the burden of losing profit through subsidizing domestic consumption is weighing on NIOC's ability to invest in the upstream sector. This is why the subsidies issue figures so prominently in Iranian political debate.

In Kuwait, the investment of reserve funds created by previous surpluses of petroleum revenue has given the government a significant stable source of revenue, equivalent to around one third of the non-hydrocarbon GDP. Assets in the Kuwait Reserve Fund for Future Generations (RFFG) and other financial reserves were estimated at the end of the financial year 2003–04 at $74.7 billion and $22.7 billion respectively (*MEES*, 5 July 2004). The figures for Kuwait in table 2 include, as do the IMF reports, income from the RFFG (which is not accessible for current government expenditure) before the mandatory contributions to it. The figures in brackets exclude these items.

The Emirates differ in their dependence on hydrocarbon revenues in their state budgets, from over 80 percent for Abu Dhabi to 40 percent for Dubai. In the UAE, information about Abu Dhabi's reserve funds (the Abu Dhabi Investment Agency) is not available to the public: its assets in 2003 were estimated at $200 billion (Bloomberg, 30 January 2005). Like Kuwait, the UAE normally shows a surplus on government current operations.

In Algeria, financial and fiscal reforms have been slow. The IMF has recommended that the Fonds de régulation des recettes (Hydrocarbon revenue stabilization fund) should be incorporated into the budget framework, with a statutory share of revenues committed to it and its use limited (IMF, *Country Report 05/05*, p. 15).

Foreign Exchange

The non-petroleum sectors of the economies of Iran, Saudi Arabia, Kuwait, the United Arab Emirates and Algeria are structurally in a balance of payments deficit on the current account. This means that the income from non-petroleum exports is insufficient to pay for imports of goods and services, interest on foreign debt and (in the Gulf Arab countries) transfers of income by expatriate workers.[13] Petroleum export earnings cover the deficit as far as they can, leaving a surplus in the general current account when revenues are high and a deficit when prices are low (current account balance = petroleum export earnings – non-petroleum deficit). The non-petroleum deficit comprises imports, payments for foreign services, interest and income paid abroad, less exports, by the non-petroleum sector. This is shown for the five countries in table 3. The level of the non-hydrocarbon current account deficit (NHCD) as a percentage of the whole economy's current account is a measure of the external dependence of the economy on petroleum export earnings. It is more meaningful than the proportion of hydrocarbon revenues in total exports. Because of volatile oil prices, current account payments are more stable than overall current balances and reflect the economy's continuing need for foreign exchange. Also, in most countries there are significant non-trade payments on the current account: remuneration for foreign investment in the hydrocarbon sector in Algeria and the UAE, payments for services in all countries, and remittances by migrant

13. In Algeria and Iran, there is a net inflow of funds from Algerians and Iranians working abroad.

Table 3. The Non-Hydrocarbon Sector in the Current Account of the Five States

Current account	*Year*	*NHCD as a percentage of total payments*	*Non-hydrocarbon current account balance ($bn)*	*Total economy current account balance ($bn)*
Algeria	2002	79	–14	4
	2003	77	–15	9
Saudi Arabia	2002	73	–48	12
	2003	72	–50	30
UAE	2002	68	–17	3
	2003	73	–20	7
Kuwait	2002	66	–10	4
	2003	70	–12	7
Iran	2002	63	–19	4
	2003	63	–25	2

Sources: Algeria, Iran, and UAE: IMF country reports; Saudi Arabia, Kuwait: central banks.

N.B.: Because there are no separate statistics for imports for the petroleum sector, these are included in the "payments" column.

workers (inward in the case of Algeria and Iran, outward in the case of the Gulf states). For the five states, the non-hydrocarbon current account deficit ranged between 60 percent and 80 percent in 2002–03. The current balances, because of structural differences, were more diverse.

The Outlook

Simulations on the Non-Hydrocarbon Fiscal Deficit

Several simulations[14] have been made to show the capacity of the hydrocarbon sector in the five states to support the rest of the economy over the next 25 to 30 years. These are based on various oil price and production scenarios, with long-term price scenarios of $25 or $35 per barrel.[15] The key factor is the oil production plateau. On present plans for expanding production (as listed in chapter 8), current official estimates of proved reserves and current depletion policies, production of liquid hydrocarbons will cease to grow between 2008 and 2012 in Algeria, around 2011 in the UAE, around 2020 in Kuwait, 2025 in Iran and—depending on the depletion policy chosen—between 2016 and 2034 in Saudi Arabia.

14. The key assumptions behind the simulations and the hydrocarbon production profiles for each country can be found in appendix 3. Only a sample of the simulations graphed is shown.

15. References to oil prices are to the average price of Brent crude in U.S. dollars in 2003.

In general, with unchanged policies, the current degree of dependence cannot be sustained much beyond the date at which production levels off. How far beyond that it can be sustained depends on the price scenario. With policy adjustments (such as capping or slowing the growth of the non-hydrocarbon fiscal deficit and phasing out subsidies), the critical date can be deferred. The figures below also show the importance of the price scenario to the time that each country has in which to reduce its dependence on hydrocarbons. The long-term price trend is also important. If the oil price remains historically high and if there were to be no consequent change in the growth of deficits in the dependent sector (which is unlikely), there is a chance to defer the crunch points until a much later date than those given in the $35/b scenario; but this may retard the development of the non-hydrocarbon sector in the long term if windfalls are not invested prudently. Likewise, there may be incentives to produce more, earlier and thereby bring about a quicker decline in reserves. However, higher prices will make the problem more difficult if government spending in the non-hydrocarbon sector adjusts to reflect the higher oil revenues—without a corresponding increase in the capacity of the non-hydrocarbon sector to generate tax revenues and foreign exchange. Moreover, in all cases, once oil production reaches a plateau, oil exports (the difference between production and consumption) will decline as consumption rises with the non-hydrocarbon GDP.

All the countries discussed in this book face the same broad challenges, but each will be vulnerable to its specific circumstances. The general warning of the simulations is that adjustments to hydrocarbon dependence must begin when production starts to plateau; they must not be delayed until it eventually falls off.

$25/b. In all the five countries except the United Arab Emirates, the base world oil price of $25/b would not sustain a five percent growth rate in the rest of the economy for long in view of its present degree of support from the hydrocarbon sector. Policy adjustments would be needed.

Saudi Arabia would not be able to support the NHFD and current domestic fuel prices (which contribute to the growth of the non-oil economy) at this price, even if the NHFD were capped from 2015 onwards (see figure A3-2, appendix 3).[16] However, the non-oil "need" can be reduced to sustainable levels if the government also brings domestic fuel prices into line with international ones by 2020. Under the conservative scenario of 260 billion barrels of

16. This is a representation of the "they'll need the money" argument that underlies the conventional forecasts that Saudi production of 15 million b/d or more can be sustained beyond 2025. They rely on early success in increasing proved reserves.

reserves, 12 million b/d production and a $25/b price, this "need" could be sustained until about 2035, when production would decline. With 330 billion barrels of reserves, it could be sustained beyond 2050.

In Algeria and Iran, hydrocarbon revenues would fail to meet the NHFD after around 2006 under this scenario if no adjustments are made. In Iran, a phasing out of subsidies by 2015 would delay the critical point to 2020; and the additional capping of the NHFD by 2025, the likely time of the oil production plateau, would ameliorate but not resolve its predicament. There would be increasing tension between the "need" for revenue from exports and the "need" for subsidized energy to consumers.

In Algeria, oil production is likely to reach its peak and decline much earlier than in the other countries: it will flatten in 2008 and begin to decline in 2012 if no substantial new discoveries are made. Increasing gas production would compensate for declining liquids production (crude oil and gas liquids) until about 2020, when gas production too will begin to decline. With annual consumption continuing to grow at five percent, Algeria could cease to be an oil exporter in the early 2020s and cease to be a hydrocarbon exporter about 10 years later. Therefore, the race is on to increase the non-hydrocarbon sector's export capacity before resources run out, and there is not much time in which to do it.

The United Arab Emirates is an exception. Although a production plateau is currently forecast for 2011, lower production, of 3 million b/d, could still sustain the NHFD until around 2010 at $25/b; higher production (with oil and liquids production rising to 6.2 million b/d by 2030) at $25/b could sustain it until around 2016. This takes no account of the investment income from the reserve fund, now managed largely by the Abu Dhabi Investment Authority. If this were released, instead of being accumulated, the deficits could be sustained for another three to five years.

The UAE will not need to make radical adjustments in the short to medium term (except in the highly unlikely case that the price of oil falls below $25/b) because of its population structure and comparatively strong non-hydrocarbon sector. Unlike in the major oil exporting countries, the UAE's (in particular Abu Dhabi's) high degree of dependence on oil and gas is sustainable for decades, even at relatively low oil prices, without major increases in production.

In Kuwait, even if production is increased to 4 million b/d by 2020 and accumulated revenue in the RFFG is run down, a price of $25/b for oil after 2006 means that the NHFD would be unsustainable after 2015. However, because of its population structure, Kuwait has the option of adjusting the needs of its non-oil economy to available revenues at the expense of its migrant population, or at least potential migrants. Under the adjusted scenario in which the annual growth rates of the NHFD and the current account deficit fall to 1.75 percent by 2010, the needs of the non-oil economy are brought roughly into line with the revenues (including investment income) available under the $25/b scenario. This implies a large shift of growth from low-cost labor activities to higher value added (and tax generating) activities, a shift in the balance of the Kuwaiti labor force to the private sector and higher investment in the capital stock with which they work.

$35/b. An average long-term oil price of $35/b would ease the adjustment challenge to each country to a different extent.

In Saudi Arabia, production of 12 million b/d at $35/b would allow something close to "unchanged policies": subsidies could be maintained and the non-hydrocarbon fiscal deficit could grow at four percent per annum until about 2025 (for projections of Saudi oil revenue based on the price of $35/b, see figure 1). There would be some scope for higher growth in expenditure and raising production to 15 million b/d. Compared to the $25/b scenario, Saudi Arabia would need higher prices or higher production, but not both, in order to maintain the dependence of the non-hydrocarbon sector on the petroleum industry.

In terms of planning, exploration and production strategy, the Saudi petroleum industry will need to take into account several factors:

—If 1.5–2 million b/d of spare capacity must be accommodated within a 12 or 15 million b/d plateau capacity, the revenue plateau will be reached sooner, and adjustments in economic policy will need to show results in a 10–20 year time frame;

—Plans to increase gas production are essential to enable oil exports (and thus oil revenues) to expand after 2010. Even with the gas strategy, oil production must be increased beyond 12 million b/d so as to make it possible for exports to grow beyond 2015. With 15 million b/d of production, exports would not expand after 2025 unless gas discoveries and developments permitted a further switching of domestic oil demand to gas.

—To sustain production at 15 million b/d, it would be necessary to increase significantly the 260 billion barrels of Saudi proved reserves by confirming probable reserves.

In Iran, the present degree of dependence on the hydrocarbon sector could be sustained until 2015, after which the contribution of petroleum would grow more slowly than the needs of the non-hydrocarbon economy. Adjustments similar to those necessary if the price of oil were $25/b would then be

Figure 1. Saudi Oil Revenue, 2010–50: Need versus Value at $35 per Barrel

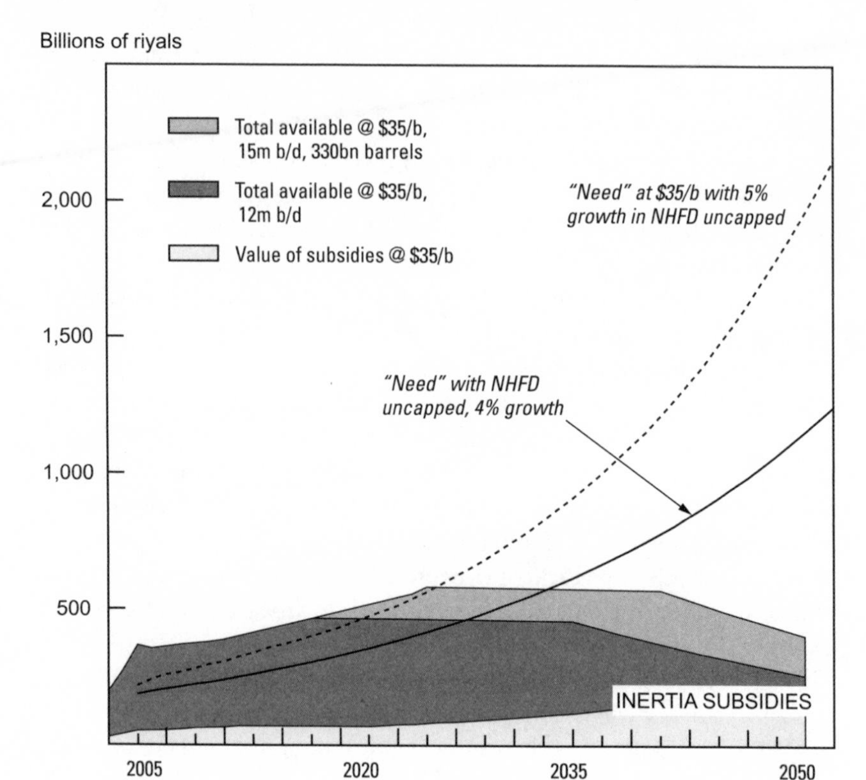

Source: Author's simulations; see appendix 3.

required in the non-hydrocarbon sector. Beyond 2025 at the latest, the maximum contribution of petroleum (at any price) would be capped by the plateau in production; but at $35/b this would, with policy adjustments, be sustainable (see figure 2).

In Algeria, an oil price of $35/b would not avoid the need for policy changes. There would be fiscal surpluses until around 2012, when hydrocarbon production would begin to decline. After that point, revenues would also decline rapidly. Roughly speaking, a 10 percent increase in today's level of proved reserves would extend the plateau of production by one to two years; but the gap between revenue and "need" would continue to increase (see figure 3 for a comparison between $25/b and $35/b revenue scenarios).

This role for the hydrocarbon industry is necessarily transitional. The length of the transition depends partly on external factors such as the devel-

Figure 2. Iran: Sustainability of Oil Revenue at $35 per Barrel, 2000–30

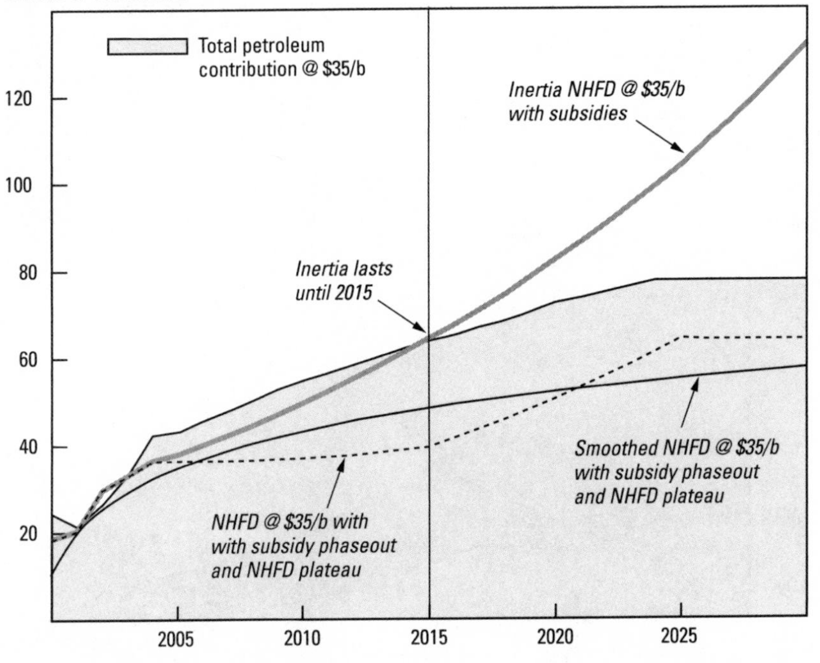

Source: Author's simulations; see appendix 3.

opment of the EU gas market. The Algerian hydrocarbon industry must match an expansion of production with an expansion of gas export infrastructure based on reasonable security of market outlets. The EU association agreement, combined with EU internal energy market reforms, requires changes in Algeria's past practice of securing long-term gas markets by exclusive agreements with importing gas companies.

In Kuwait, the $35/b scenario would, roughly speaking, avoid the need for adjustment until after 2035, allowing policies to remain unchanged (or even to increase the non-hydrocarbon sector's dependence on oil). This contrasts with the adjustment that would be necessary under the $25/b scenario (see figure 4).

In the United Arab Emirates, low production at $35/b (with a plateau of 3.6 million b/d of oil and liquids from 2010), as with high production at $25/b, would keep the NHFD going until 2016 (see figure 5). Thereafter, a gap would rapidly appear between revenue and deficit, although this could be met first with money from the reserve fund.

Figure 3. Algeria: Oil Revenue at $25 and $35 per Barrel, 2000–30

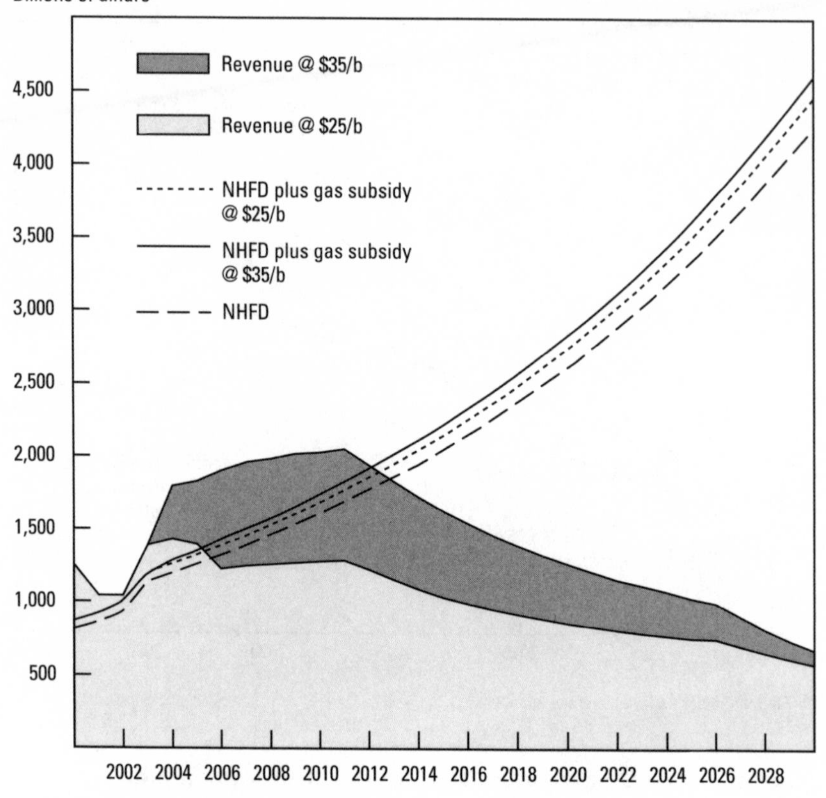

Source: Author's simulations; see appendix 3.

Simulations of the Balance of Payments

Simulations of the future evolution of the current accounts of the five hydrocarbon exporting countries demonstrate the difficulty of maintaining the current dependence on hydrocarbon revenues for funding imports. Assuming that the non-hydrocarbon current account deficit grows with the non-hydrocarbon sector and given that oil/gas exports will at some point decline (as domestic consumption increases and production falls), there will come a critical point. After this point, it will not be possible for the income from hydrocarbon exports to fund the NHCD unless adjustments in the non-hydrocarbon sector have altered the relative levels of imports and exports sig-

Figure 4. Kuwait: Oil Revenues and Adjusted Debt at $25 and $35 per Barrel, 2000–35

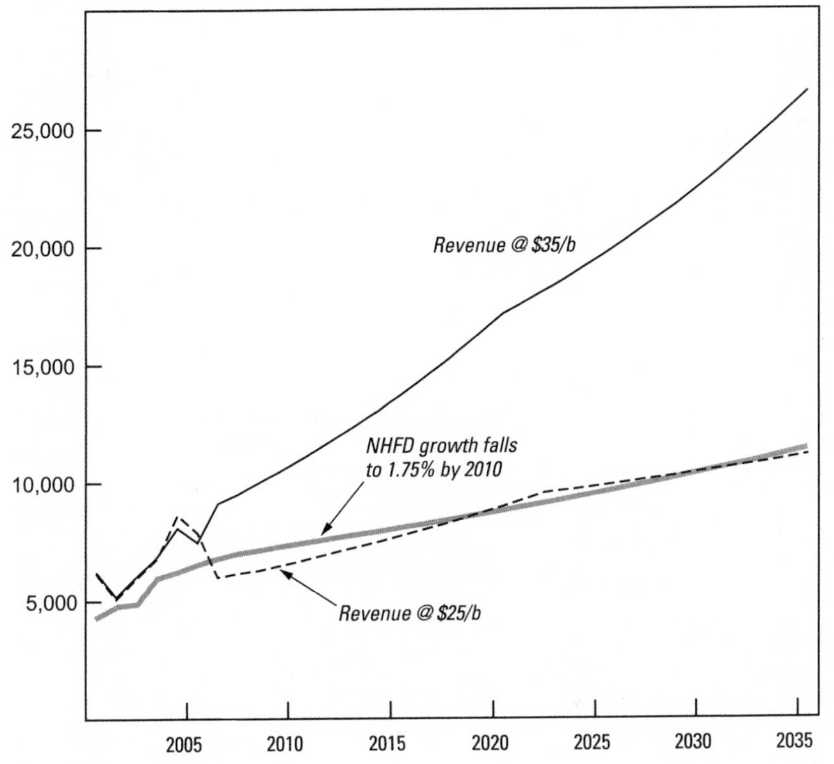

Source: Author's simulations; see appendix 3.

nificantly. All hydrocarbon exporting states risk reaching this critical stage, similar to the fiscal deficit crunch point.

Though the fiscal and export crunch points are both related to the time when production volumes level out, there are different variables. The inexorable rise of domestic consumption will usually begin reducing export volumes even before production levels out. On the other hand, export earnings per barrel are generally higher than government revenues per barrel (because export prices cover costs). The export capacity of the non-hydrocarbon sector is also a factor: in these simulations, it is assumed simply to keep pace with the growth of the non-hydrocarbon economy. In reality, reducing dependence

Figure 5. United Arab Emirates: Oil Revenue at $25 and $35 per Barrel, 2000–30

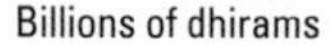

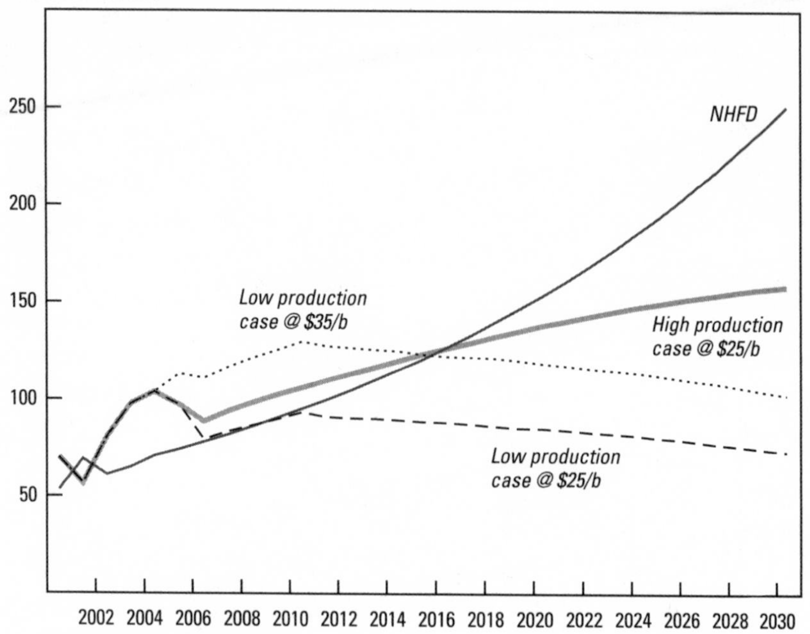

Source: Author's simulations; see appendix 3.

requires the non-hydrocarbon sector to improve its capacity to export. Finally, overseas financial assets (particularly in the case of Kuwait and the UAE) may be drawn down for some years in order to cover the current account deficit.

Even though a current account deficit can be sustained for lengthy periods of time, once the critical point is reached, pressure for change will mount. Because the deficit will be structural (owing to diminishing revenue from exports in relation to growing demand for imports) rather than cyclical (linked simply to the current economic phase), waiting for conditions to change will not be an option. These countries will be consuming more than they can afford to pay for from current income.

Some policy adjustments are possible that could reduce current account hydrocarbon dependence and avoid, or delay, the evolution of an unsupportable current account. First are those that will not alleviate hydrocarbon dependence but will delay the crunch point. These adjustments include:

—The elimination of energy subsidies to the domestic economy. This would reduce the volume of domestic hydrocarbon consumption and therefore free up greater volumes for export;

—Higher oil prices. These would raise export revenue and delay the critical point: an oil price of $35/b would hold off a structural payments deficit by between five and ten years, depending on the country; and

—Raising hydrocarbon production volumes. Rises would provide a clear means to support the increase of imports, transfer payments and investment income paid abroad. All the countries in this study have plans to increase production capacity, as discussed in chapter 8. The impact of increased production on price would obviously need to be considered. This is particularly important for Saudi Arabia. There is also a trade-off between the level of production achieved and the length of time over which it could be maintained. Thus even though higher production may fund imports now, it brings closer the critical point when non-petroleum imports cannot be paid for by hydrocarbon exports.

In the longer term, other measures will be needed to ensure a structural balance of payments equilibrium. These might include:

—Reducing the import propensity of the non-hydrocarbon sector and boosting its export capacity. This measure would directly improve states' balance of payments position;

—Reducing the amount of money sent abroad by expatriate workers. In some states (especially Kuwait and the UAE), there is scope for this simply by limiting the number of these workers. However, the effect on the competitiveness of the non-hydrocarbon sector would need to be considered; and

—Investment of surplus hydrocarbon revenues abroad. All countries do this, but only Kuwait and Abu Dhabi do it automatically through investment funds.

Although petroleum export dependence is common among the countries under consideration, there are some differences in their balance of payments profiles. Kuwait's current account benefits from investment income (mainly from the state's reserve funds), which covered 20 percent of payments in 2003. Iran's current account depends less on petroleum than that of the other four countries. In 2003, the non-petroleum deficit was 63 percent of payments, reflecting Iran's more diversified domestic economy. The UAE has a healthy re-export sector, meaning that non-petroleum exports are a significant source of support to the current account. It has the potential to avoid serious adjustments (though they may still be desirable) for the next 30 years.

Support from this source is missing in many of the other Middle Eastern producers' economies. These contrasting circumstances, in addition to differences in reserves and production capacities, explain the differing time frames that states face within which to deal with the hydrocarbon dependence of their balance of payments. Algeria, for example, has significant changes to make if it is to avoid facing balance of payments pressure within the next 10 years. The Algerian problem (illustrated in figure 6) is most acute because of the very low export capacity (about five percent of output) of its non-hydrocarbon sector. The current account deficits of the non-hydrocarbon sector cannot be sustained much beyond 2012 at $25/b or 2014 at $35/b. Additional reserves would delay the decline from the production plateau under either price scenario. The need to develop Algeria's non-hydrocarbon export capacity is critical, and there is not much time to do it.

General Conclusions

In the 30 years since the first oil shock flooded Saudi Arabia, Iran, Algeria, Kuwait and the United Arab Emirates with money, their situation has changed in many ways:

—Their physical infrastructure has been transformed;

—Their populations have increased. Nationals are better educated and healthier, and they live longer; and

—Government sectors have expanded. Though government growth is now being curtailed, the reforms needed to advance growth in the private sector have not yet begun to expand non-petroleum tax and export revenue and thus reduce dependence on petroleum.

In the next 30 years or less, these countries will have to adapt to a life that is less dependent on oil than it is now. For some countries, adaptation would need to begin, or resume, immediately if the oil price falls back toward a $25/b average. It is arguable that the exporting countries have become committed to levels of consumption of goods and services that are "too high." The World Development Indicators published by the World Bank show these countries to be depleting their capital stock (petroleum reserves), without replacing them with sufficient other assets.[17] Economists at the IMF, following a different methodology ("permanent income"), also point to the unsustainability of the fiscal deficits of the oil exporting countries (Davis, 2003),

17. The methodology used for the World Development Indicators is explained in Hamilton 2002 (see his table A1).

Figure 6. Algeria's Current Account, 2000–24

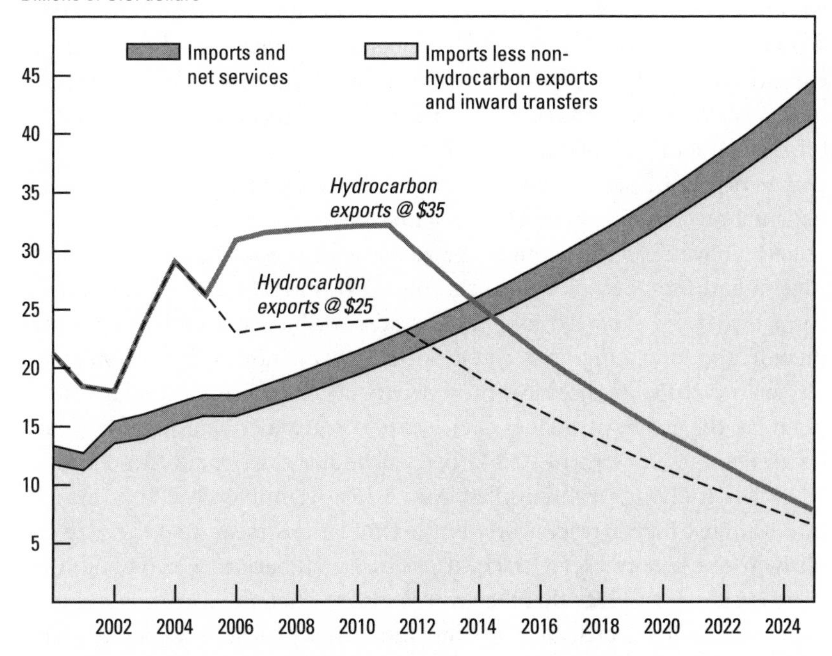

Source: Author's simulations; see appendix 3.

and this has been illustrated in IMF Staff Reports on individual countries. There is difficulty in valuing "permanent income" from unpredictable oil prices and production in such studies. Even so, our simple simulation analysis supports the idea that these countries are consuming resources at a rate that they cannot sustain without very challenging adjustments to reduce the fiscal and current account deficits of their non-hydrocarbon sectors. The problem is that the countries concerned have to start from where they are, even if, in theoretical terms, it would have been better not to have reached such levels of consumption.

There is therefore a bridging role, in most countries, for an expansion of the hydrocarbon sector in the medium term; but for the first time in a quarter century or more, it is production potential, not markets, which is likely to limit that expansion. It will be a shaky bridge. Our analysis has used two price scenarios: $35/b Brent equivalent after 2006 and $25/b Brent equivalent after 2006. Prices were treated as flat (in constant dollars), as an average price over a number of years, though, in fact, shocks and cycles are likely. Stabilization funds and financial reserves such as those of Kuwait and the UAE can counter this problem to some extent, but volatility will continue to be an acute problem for oil-dependent economies. The analysis in this chapter has emphasized the crucial need for the exporting countries to have set their fiscal and current account dependence on hydrocarbon revenues on a downward trajectory by the time petroleum revenues reach a plateau. This time is not far off, though it differs for each country.

The degree of urgency to adjust dependence on the hydrocarbon sector depends on what happens to prices in the short term. The longer prices remain above $35/b, the longer the adjustment can be postponed. But for Algeria and Iran, even a trend price of $35/b (in real terms) would not be enough to avoid the need for a combination of increasing oil and gas production and increasing taxes and exports from the non-hydrocarbon sector well before 2015. Saudi Arabia must eventually embark on a similar course, but it has the luxury of a wider choice of future production profiles by which it may delay its adjustment. At $35/b, it will be able to defer adjustments until 2025 at 12 million b/d and until at least 2035 at 15 million b/d. It is implicit, however, that higher prices early in the simulations mean that they are less likely to rise later on. Conversely, if prices do fall back to $25/b after 2006, there is more possibility that they will rise later, when the scenarios' volumes may have reached a plateau. The simulations thus present a deliberately simplified choice to governments, which have to make decisions about expanding capacity—and therefore about the investment programs of the NOCs.

Market uncertainties are great. After the 1990s and the early years of the present decade, in which prices slumped dramatically, it was understandable for exporting governments to follow a "wait and see" policy. Even though times have looked up for oil producers in 2004–05, it remains feasible to construct a scenario in which the impact on the world economy of the very high prices at this time, slackening demand for oil from China and increasing production from Russia, Iraq and elsewhere could create a more relaxed market in the medium term. At this stage, export revenues could fall toward the $25/b level. Even if the market accommodates some increase in volume from the countries in this study that have the capacity to produce it, this medium term will be the time when the weight of the non-hydrocarbon sectors will really begin to burden the hydrocarbon sector if these countries' degree of self-sufficiency is not increased.

With continuing dependence on the hydrocarbon sector, growth in the non-petroleum economy increases domestic consumption, reduces export volumes and tends to reduce government revenues. Faced with this possibility, it makes sense for the exporters to increase petroleum capacity as a defensive hedge and to hope that by cooperation in OPEC and with other exporters, they manage to avoid a damaging competition led by price. For the non-petroleum economy, levels of long-term support available from the hydrocarbon sector are unpredictable within a wide range. The exceptions are Kuwait and the UAE, where the size of overseas investments in relation to the population is large enough to reduce significantly the impact of this uncertainty.

The effect of uncertainty about the level of support from the hydrocarbon sector can be reduced by developments in the tax and export capacities of the non-hydrocarbon sector, ahead of what might be required under a "high-price" scenario and consistent with the "low-price" scenario under which they would be unavoidable. If these are based on long-run competitive, export-oriented, profitable businesses, the countries cannot lose.

appendix one
Questionnaire

Participants usually received the following guidelines for discussion in writing, ahead of time. When appropriate, they were discussed at the interview:

—Interviews are anonymous and not recorded;

—The more sensitive information is for your company's use only;

—You do not have to answer;

—The purpose of the study is to get insight from the key Middle Eastern oil producers on how they see the future challenges to their industry.

In the book, interviewees are identified with a code (for example, SA16 or IR3) in order to preserve their anonymity.

This questionnaire presents the main questions asked. They covered a wide range of topics. I chose the most appropriate questions on the basis of the interviewee's expertise and position, and also pursued other areas of inquiry, depending on interest demonstrated during the conversation. The phrases in brackets indicate potential suggestions I might introduce to provoke an answer if none were spontaneously forthcoming. Additional questions were designed for specific country situations. Some of them are reproduced after the general questionnaire. Specific questions were also designed for young professionals.

The first question was my lead question in almost all cases, save in Iran, where it did not have sufficient resonance.

—Would you describe your company as international or national? Is that going to change?

Mission

—Why was your industry nationalized?
—How have circumstances changed since then?

—What is the mission of your national oil company (rent collector, employer, center of technology)?
—How far does your mission go beyond business into, say, providing jobs, local procurements, social services?

—Do you see a transformation of your core business from the exploration and extraction of crude to marketing the refined product?

Corporate Culture

—How important is it that the state controls the oil?
—Is there any chance of that changing?

—Do you think that the political authorities (the oil ministry, for example) would resist the stronger development of a corporate culture? A stronger *esprit de corps* in the NOC?
—Do you think the younger generations in the company have different mentalities from the generation of their fathers?

—What is your company's greatest accomplishment?

—Is there a future for Islamic project financing in the oil and gas sector?

—Do you think national oil companies in the Middle East should develop their own business style? Or should they strive to have the best Western-style business?
—What is the Western business style?
—What could be the Eastern business style?

Future Challenges to the Industry

—What are the greatest challenges to your oil industry in the next 10 years?

—Are the challenges greater internally (need to develop capacity, unemployment) or externally (market share, non-OPEC, demand)?

—How important is the market to what you do? How high does it figure in your discussions about planning or performance?

—Do you feel that you influence the market or that you are dependent on it?

—What does the company lack to fully develop its potential? [I show the following list:]

- Technology
- Investment capital
- Management practices
- Marketing capacity
- Autonomy from government
- Price stability, price control
- Other
- None of these

External Drivers

—Do you think a U.S.-controlled Iraq will drain the investment capital available in the region?

—Could competition from Iraq push your government to offer better terms or more opportunities to foreign investors?

—Would you say the challenge from Russian oil and gas is a more serious threat to your market share and your influence on the international markets?

—Do you think that developing joint ventures with Russian partners could help to counter that threat?

—Would your country support an OPEC policy aimed at regaining the market share lost to non-OPEC oil?

—How long could you sustain low prices? Or a price collapse?

—Do you think it is too late for OPEC to regain its market share lost to Russia and the Atlantic basin? (Wouldn't those producers weather low prices now that investments are made?)

Marketing (Finance, Refining, Marketing, International)

—Would you consider that setting the price for an export contract of refined products is now a routine commercial decision for your company? Or does it involve special decisionmaking?

—How is the pricing methodology different for crude oil?

—Is this decision on price taken outside the company?

—Are you looking into the use of futures and derivatives markets as a way of hedging the NOC's expected cash flows?

—How do you work around the constraints imposed by OPEC oil policy?

(How does OPEC policy affect setting prices, marketing volumes and different products?)

—Is price volatility a preoccupation for you in your job or for your company?

—Who are your competitors in the market (NOCs, IOCs, traders, financial institutions)?

—Who are your partners in the market (NOCs, IOCs, traders, financial institutions)?

—When I read *The Prize*, I was struck by how the oil industry was built by oil companies first extracting crude, then transporting it, then refining it—basically taking on the activities all through the chain, integrating activities to control the price of their crude and its destination. Now these companies are still very integrated, but they increasingly abandon the need for control of their production through to the end-user. Oil is now seen as more fungible. Meanwhile, NOCs are following a different path. They are integrating, and some are internationalizing their activities; but they maintain a desire to control price formation and to control their molecules throughout the chain. Do you think this is an accurate assessment?

—Why do NOCs focus on control?

—Do you think this will change?

Sustainability of the System (Economy)

—Do you think people in your country are more concerned about politics or economics?

—In view of the increasing needs of the population and the government turning to the NOC for revenues, do you think the NOC will be less and less independent of government?
—Or will it become increasingly important and powerful?

—Will the NOC be able to sustain the economy over the next 10 to 20 years as it has done in the past?

—Have you seen any changes in the demands made on the NOC by government?

Role of Government in Determining Strategy

—What directives do you receive from the government for oil policy (expansion of capacity, surplus capacity, maintaining market share, OPEC price policy)?
—Are these decisions the result of discussions with NOC experts or top managers?

—What do you think of the triangular management/regulation model? [I draw the model and explain it.]

—Does the NOC retain its profits and pay taxes or does it surrender all its cash flow and get approval for individual projects?

Corporate Strategy

—How do you determine strategy?

—What is the role of the Supreme Petroleum Council, the executive committee?
—Who sits on the petroleum council and the executive committee?

—Would you like to see the NOC have less influence on determining oil policy if it meant more freedom to manage its business activities?

—Do you discuss oil policy internally?
—Do you meet with the oil ministry to discuss policy?

—How far ahead do you plan (5, 10, 20 years)?
—Would you say that you usually follow the plans you set out?
—What diverts you from the plans (volatility of oil price, government needs for extra revenue, unavailability of capital)?

—What is your company's strategy for increasing production revenues (reserves replacement, enhanced recovery)?

—Do you see your company developing internationally?
—Is that an important part of your company's outlook for the future?

—Why internationalize your business?
—Is it a strategic decision to secure outlets for your crude (like Pemex and PDVSA did in the United States)? Or is it to expand your business upstream (like Statoil did to offset declining domestic reserves)?

—Does the international side of the business escape somewhat from the government's control?

—What are the obstacles to internationalizing the business?

—How important is the development of gas in your strategy?
—Is your gas production mostly flared or non-associated?

Role of/Relations with IOCs

—What has been your experience of technical cooperation with IOCs?
—What worked well? What worked badly?

—Do contracting companies (Bechtel, Halliburton, Schlumberger) provide similar or better access to technology than IOCs?
—How does what they offer compare to that offered by IOCs?

—How do you think the company would respond if the political authorities wanted to give a foreign oil company a concession to explore and develop oil?
—Is it different for gas?

—(If there is foreign investment) Is it more likely that IOCs would participate in existing developments or new developments of known reservoirs or that they would get licenses for new exploration?

—Is there a preferred type of investor for each one of these kinds of activity (IOCs for exploration, smaller companies for smaller risk production and service companies for enhanced recovery)?

—What can international oil companies offer?

—Are they all alike?

—Would your company gain more from, say, Statoil (an ex-NOC), BP (a super-major), Total (a more flexible negotiator) or an American company supported by Washington?

—Which IOC has the worst reputation? Why?

—Which has the best? Why?

—Do you think your company would benefit from a joint venture with that IOC?

—What would the perfect deal with an IOC look like?

—Do you think IOCs want to invest in oil or gas in your country no matter what social or economic impact they might have? In other words, how important is a win–win deal to the IOCs?

—Which company would you choose to do business with?

—Would the oil ministry feel differently?

NOCs and Other Players

—What do you think of working with NOCs from other countries and in particular NOCs from developing countries (CNPC, Petronas)?

—Do you have any experience of working with them?

—Do they bring anything different?

—Are they safer partners?

—Asian oil companies, both private and public, have shown a great deal of interest in investing in the upstream and downstream oil business in the Middle East. What do you think of them as partners?

—Do they have advantages over Western companies?

—What do you think of the Russian companies that are trying to internationalize their businesses?

—Which country in the region is most likely to invite foreign investors in oil, in gas?

—Are there any regional companies with whom your company could do business?

—Do you think that foreign investment taking the shape of joint ventures between IOCs and developing world NOCs would be easier to sell politically in your country?

Additional Questions in Iran

—Some observers have warned that production capacity could fall to 3 million b/d without the development of new fields. How steep are the decline rates?
—How do you plan to offset them (gas injection schemes, foreign capital)?

—Is NIOC pleased with the participation of the IOCs so far? Is there a feeling that your industry has benefited?

—What is your impression of the history of working with the consortium before complete nationalization?

—According to oil minister Bijan Zanganeh, Iran's strategy is to maintain its production share within OPEC. Does Iran want spare capacity? Why?

—How do you manage an expansion of production within the constraints of the government's OPEC policy?

—There have been many helpful statements of support made toward Iraq and its oil industry. How do you foresee increased cooperation with Iraq?
—What obstacles are there?

—Is regional cooperation a high priority?
—Which regional partners seem most likely?

Additional Questions in Algeria

—Comment les relations entre Sonatrach et les compagnies étrangères ont-elles évolué depuis la nationalisation?

—Aujourd'hui y a-t-il une attitude différente face aux compagnies françaises vis-à-vis des autres compagnies étrangères?

—Que s'est-il passé depuis 2000 lorsque le Ministre Khelil a proposé la réforme du secteur? Où en sommes-nous aujourd'hui?

—Les réformes proposées par le Ministre Chakib Khelil vont-elles trop loin, trop vite, ou sont-elles vraiment ce dont le secteur énergétique a besoin en Algérie?

—Que pensent les gens de Sonatrach des réformes?
—Y a-t-il une différence d'attitude entre les générations?

—Sonatrach a mené une politique d'internationalisation par laquelle elle cherche à acquérir des parts minoritaires dans toutes sortes de projets pour accroître son expérience. Quel succès a eu cette politique ?

Questionnaire for Young Professionals

—Do you think that international experience is important for your career?
—Have any of your colleagues been trained abroad?
—Are they different when they return? (Do they have different attitudes?)

—Do you think you know more about the West than your father?
—What about Western business practices?

—Does everyone in your team speak English?

—Do you think your country's educational system trains young professionals as well as Western universities do?
—How does it compare? (How is it different?)

—Would you like to see your company developing stronger professional ties with outside partners?
—Do you favor IOCs?
—Do you favor regional partners?
—Do you favor Islamic project financing?

—How important are family ties in the business (doing favors for family and friends, giving business opportunities to people you know rather than those you do not)?
—Is this changing with the younger generation?

—Would you like to see your company become more Western managed?
—What is the difference?

—Why was your oil industry nationalized?
—Are those reasons still as strong today?

—Would you approve of a partial privatization of your national company if that made it more competitive?

—Do you think there is a strong sense of pride in the company?

—Do employees feel that the government inhibits or supports your work?

—Do you think the government wastes the oil revenue?

appendix two
National Oil Company Structures

Abu Dhabi National Oil Company and Subsidiaries

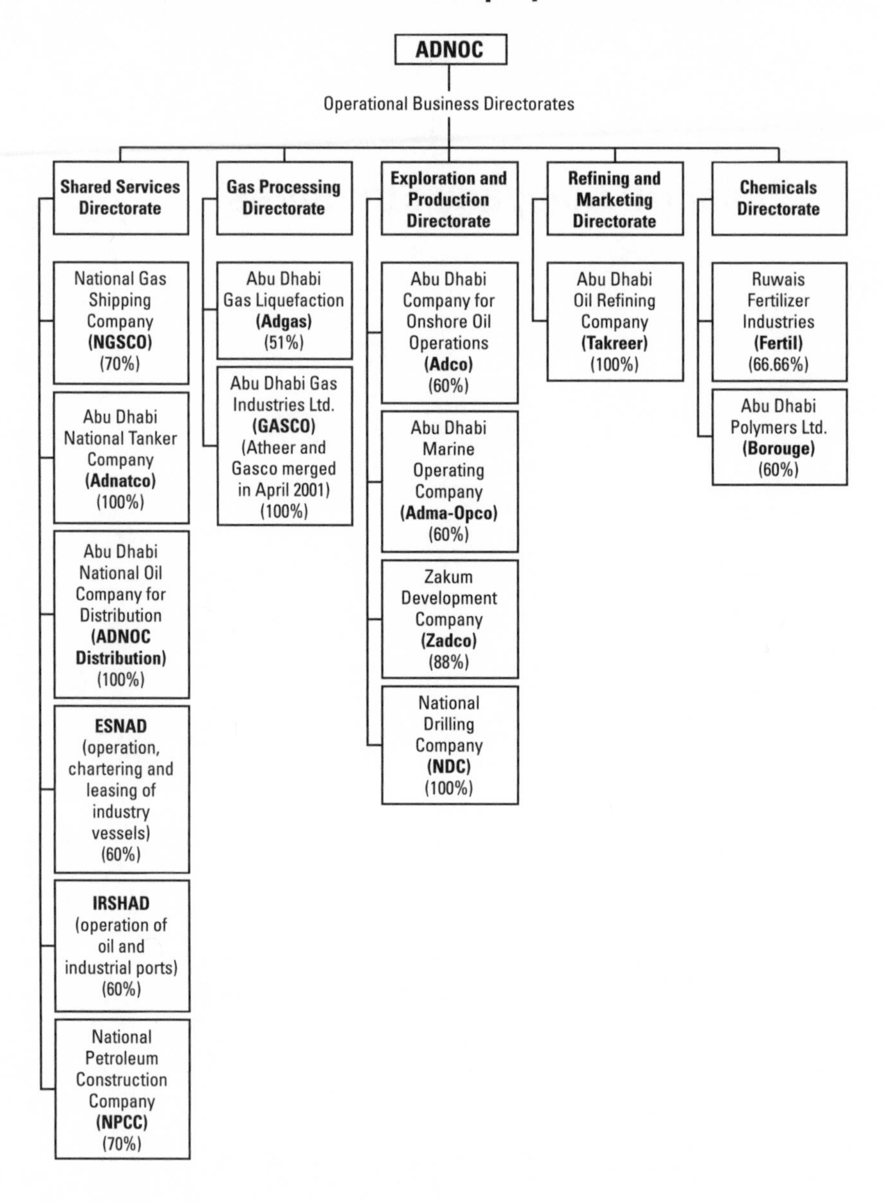

Kuwait Petroleum Corporation and Subsidiaries

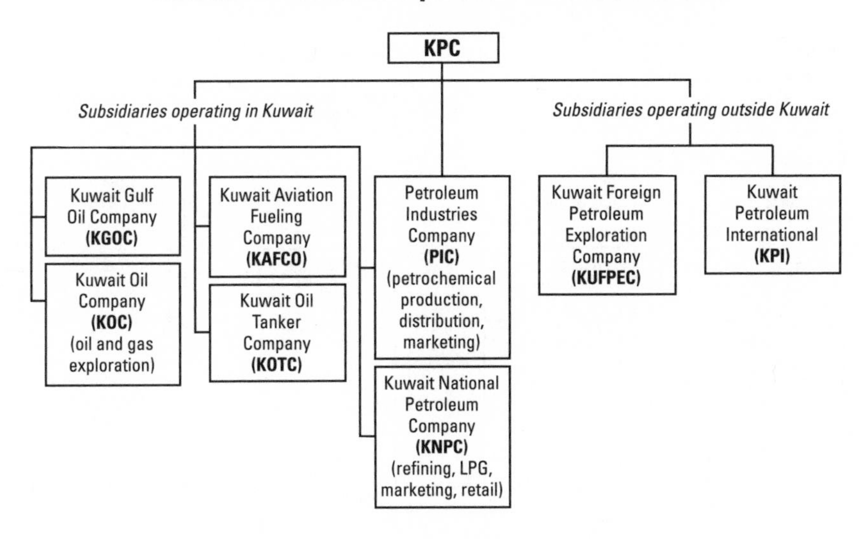

National Iranian Oil Company, Its Sister NOCs and Subsidiaries

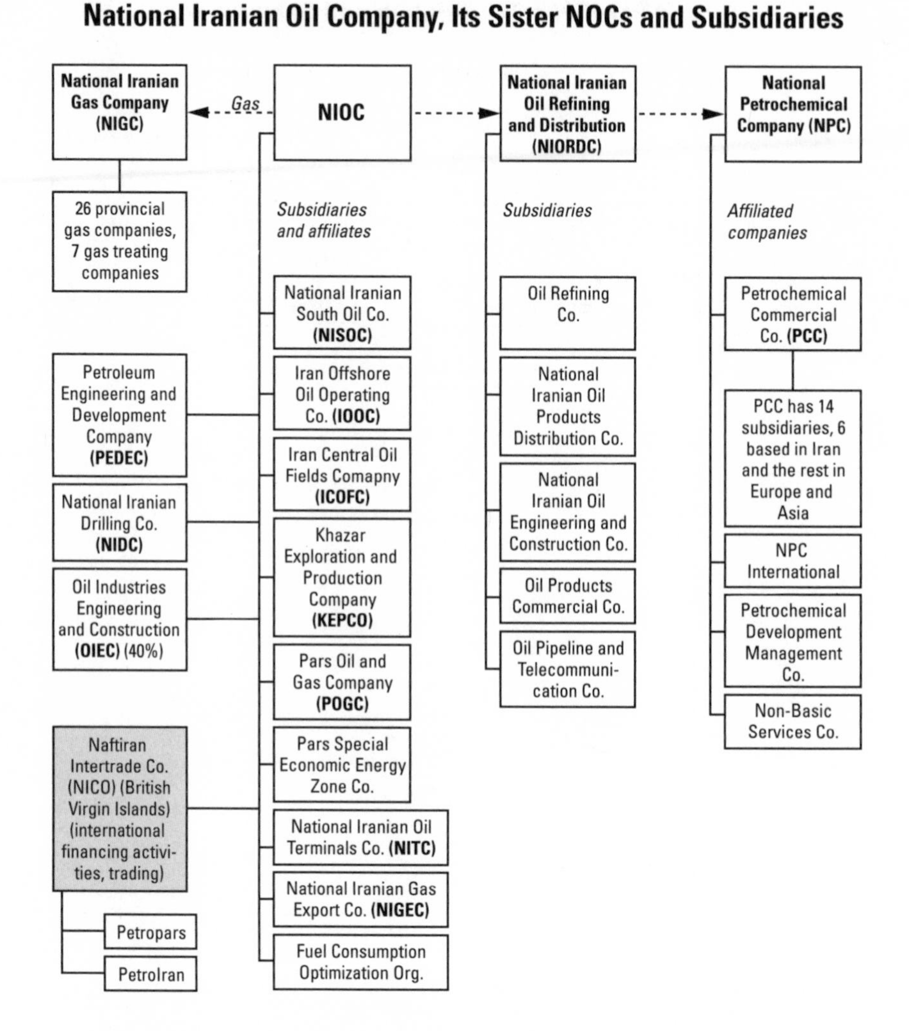

Saudi Aramco and Subsidiaries

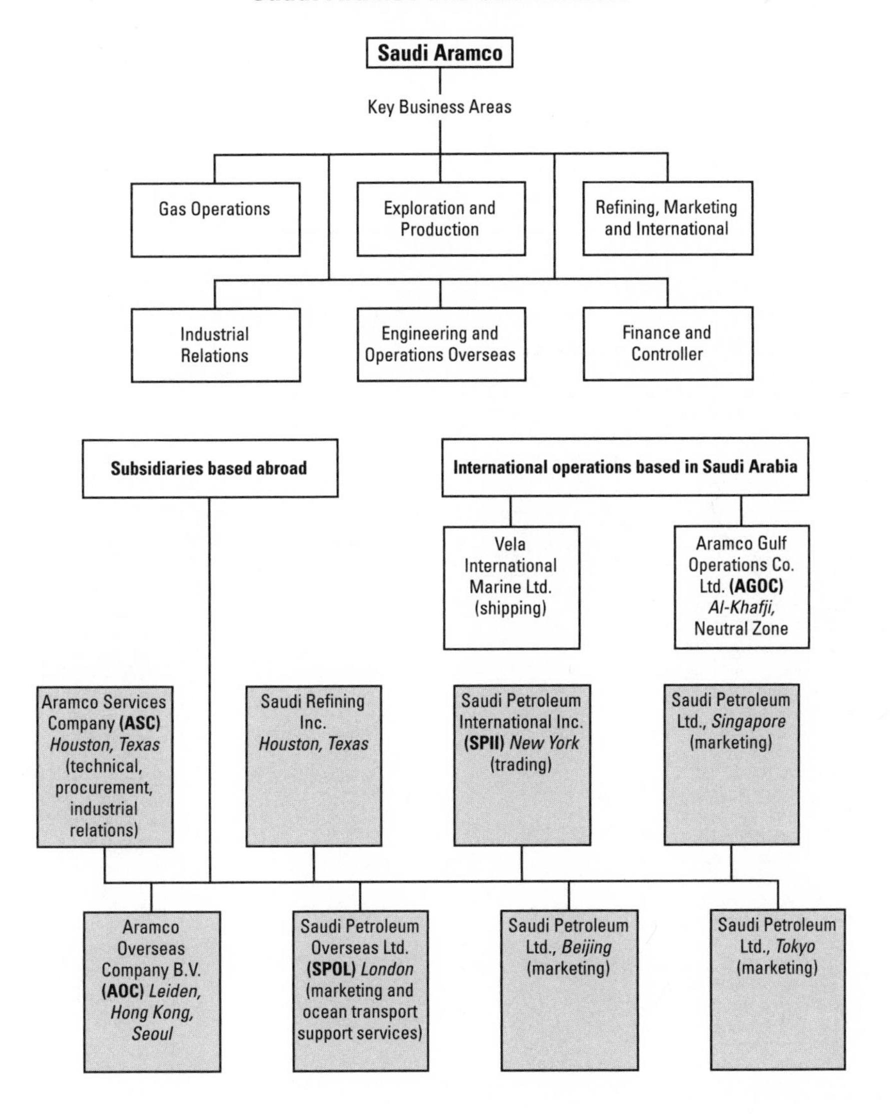

Sonatrach and Subsidiaries

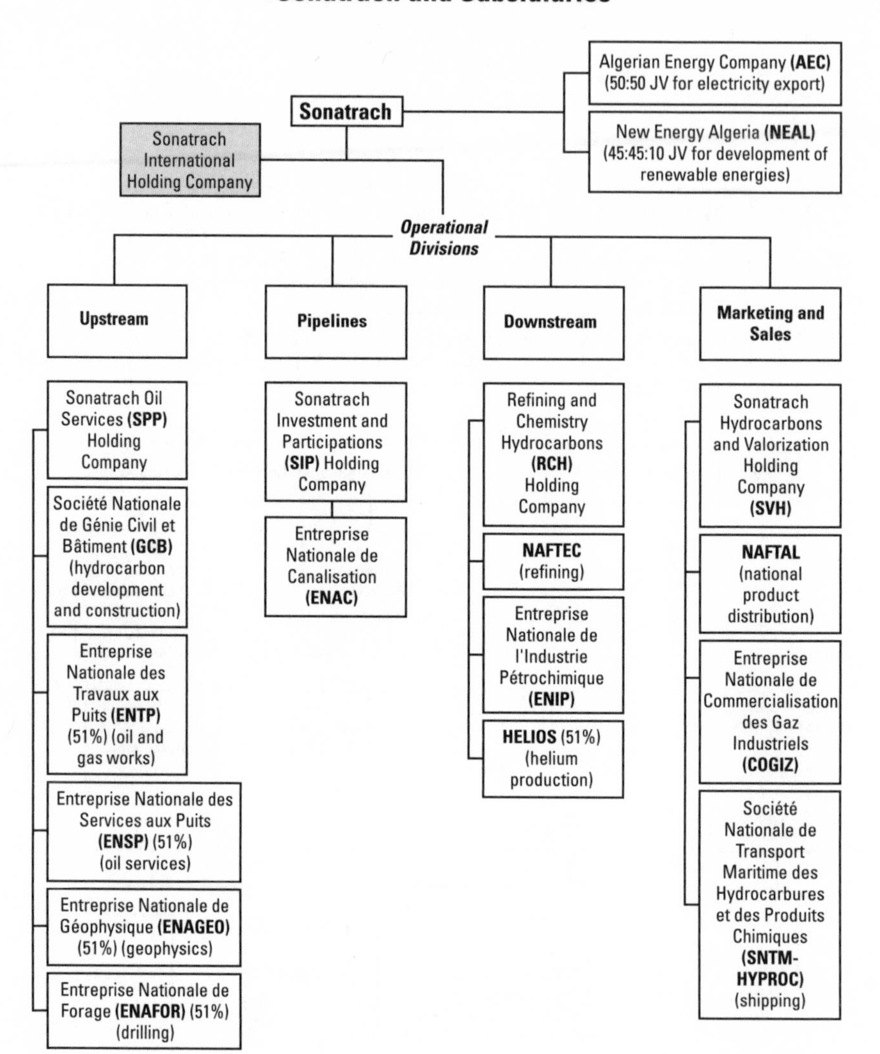

appendix three

The Challenges of a Petroleum-Dependent Economy: Simulations of the Future

The simulations referred to in John Mitchell's economic background contribution were designed to show, under various oil price and production scenarios, the capacity of the government oil revenues of Saudi Arabia, Iran, Algeria, Kuwait and the United Arab Emirates to support continuing fiscal and current account deficits in the non-hydrocarbon sector of the economy. This appendix explains more fully the data, logic and outcome of the simulations.

The main conceptual challenge was to draw an artificially hard line between the hydrocarbon and non-hydrocarbon sectors of the economy in order to examine the dependence of the latter on the former under various scenarios for oil and gas production and price. The ratio of the fiscal and current account deficits to the non-hydrocarbon GDP is taken as the critical measure of dependence instead of the ratio of hydrocarbon revenue and exports to government revenue and exports in total. It is in the non-hydrocarbon sector that most of the population lives, works and uses government services.

Simulations of future trends need to be interpreted with care. Many interactions and feedbacks are omitted here, and precise numbers are used to represent loose trends. Also, the power of compound interest over a long period overwhelms other information and magnifies inconsistencies. This is particularly true in projections for Kuwait because of the large financial assets of the Reserve Fund for Future Generations and the General Reserve Fund, which absorb surpluses and deficits. In the UAE, the trend of the (growth or

Table A3-1. Indicators of the Growth of the Non-Hydrocarbon Economies of the Five States

Country	*Assumed percentage growth of non-hydrocarbon sector*	*Population growth, 1990–2003*	*Assumed population growth, 2003–15*
Algeria	5.0	2.3	1.7
Iran	5.0	1.5	1.3
Kuwait	5.0	0.9	1.9
Saudi Arabia	4.0	2.7	2.6
UAE	5.0	6.3	–0.7

Sources for population growth: World Bank, *World Development Indicators 2005*, Table 2.1 except for Algeria, for which see UN Human Settlements Programme, data page, 6.12.99.

otherwise of the mainly expatriate) population depends on uncertain future policy rather than normal demographics.

Economic Terms and Assumptions

Growth

An essential assumption is the rate of growth of the non-hydrocarbon economy in each country from 2004 onwards. The growth percentages used in the simulations are shown in table A3-1, along with the rates of population growth to which they are related. The assumed growth rates are based on the World Bank's *World Development Indicators 2005* (for 2004) or on government targets for growth if they have been made explicit. In the simulations, this growth rate is also the growth rate assumed for domestic hydrocarbon consumption and for the growth of the non-hydrocarbon fiscal and current account deficits. Therefore, the propensities of the non-hydrocarbon economy to consume energy, pay taxes and generate exports and imports are assumed not to change.

Non-Hydrocarbon Fiscal Deficit

This is the difference between non-hydrocarbon revenue and non-hydrocarbon expenditure in the central government's financial operations as reported in the IMF country reviews (which include "off-budget" items) or the central bank (in Saudi Arabia, this does not include "off-budget" items; they are assumed to be constant). As far as possible, the revenues and expenditure of the national oil companies are shown on a net basis in the central government's finances.

Non-Hydrocarbon Current Account Deficit

This is the difference between (a) imports of goods and services, transfer payments and investment income paid abroad and (b) non-hydrocarbon exports of goods and services, inward transfers and investment income received. It has generally not been possible to exclude imports to the hydrocarbon sector, but investment income and payments related to foreign investment in the hydrocarbon sectors of Algeria, Iran and the UAE are excluded.

Non-Hydrocarbon GDP

This is as reported in the IMF country reports for all the countries except Saudi Arabia (for which there is no comparable report). Data of the Saudi Arabian Monetary Agency (the central bank) are used instead.

Inflation

Actual data are used for 2000–03, and data for 2002–03 (the Arabian and Iranian years) are used as the base for future projections. In effect, the simulations are in constant 2002–03 prices—dollars for external transactions and national currencies for internal transactions.

The Hydrocarbon Sector: Terms and Assumptions

Gas and Oil Equivalence

Gas is converted to oil equivalence on a thermal basis for quantities. Gas re-injection, wastage and transformation use, and recovery of liquefied petroleum gas (LPG) and natural gas liquids (NGLs) in Algeria, Iran and the UAE follow current equivalence ratios for the country concerned or official projections where available. LPGs and NGLs are included with crude oil in export and revenue calculations. It is assumed that 90 percent of re-injected gas is eventually recoverable.

Prices

Assumptions about the international oil price are shown in figure A3-1 for Brent crude. Export f.o.b. values per barrel are assumed to differ from the Brent price (because of quality differences and freight costs) by the average of such differences between 2000 and 2003. Government revenue per barrel is lower by the average difference between export realizations and government revenue as reported in IMF reports and national statistics from 2000 to 2003.

Figure A3-1. Assumptions about the International Oil Price, 2000–35[a]

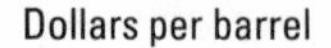

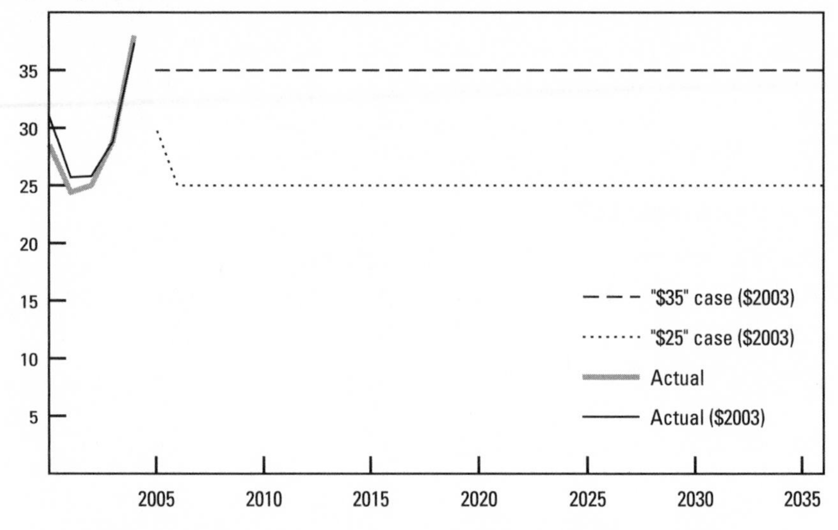

Source: 2000–04: BP: 2005, p. 14; BP: 2004, 35; John V. Mitchell's assumptions for the simulations.
a. The price is for Brent crude.

Gas and LNG price assumptions are generally based on 2003 prices or average 2000–03 prices adjusted to bear the same ratio to international oil prices as in those base periods. Domestic prices are assumed to yield no government revenue except where this is shown in IMF reports or national financial statements.

Production

Production profiles are illustrated in figures A3-2 to A3-7 below, and are compared to estimated consumption generated by the economic assumptions. The level of production at the plateau is based on recent statements from officials of the government or the national oil company. In the case of gas production in Iran and Algeria, these levels are contingent on success in signing contracts for export and involving foreign investment.

Plateau production is continued up to the point at which the current official depletion constraint (the percentage of production compared to remaining reserves) is reached, after which production declines so as to meet this constraint against the reserves. These constraints are a maximum 10 percent depletion rate in Algeria and a maximum 3 percent depletion rate in Saudi Arabia, Iran, Kuwait and Abu Dhabi. It is these latter constraints that result in production profiles lower than those extrapolated in the International Energy Agency's *World Energy Outlook* and the U.S. Energy Information Agency's *International Energy Outlook.*

The reserves used to generate constraints are the 2004 proved reserves estimates as reproduced in the *BP Statistical Review of World Energy 2005,* except where explicit allowance is made for proving additional reserves.

Notes on country production profiles, compared with the consumption generated from the economic simulations, are shown below.

Production Profiles

Saudi Arabia

This simulation is based on two alternative assumptions about Saudi oil reserves:

1. The kingdom's declared proven reserves of 260 billion barrels of oil.

2. 330 billion barrels of reserves, which include the proven reserves and 70 percent of the 103 billion barrels of reserves that Saudi Aramco has stated are the sum total of "probable" and "possible" reserves under the Society of Petroleum Engineers' definitions.[1] These figures signify that there is a 50 percent probability that total recoverable reserves exceed 292 billion barrels and a 10 percent probability that they exceed 363 billion barrels, before taking account of completely new discoveries.

The model applies production profiles that reach 10 million b/d in 2010, 12 million b/d in 2016 and 15 million b/d in 2024. Production is then continued until the reserves to production (R/P) ratio is 30:1, after which production continues at 3 percent per year of the diminishing balance. This is a conservative depletion rate, but it may be conditioned by the challenge of managing the oil-water interface on the large Ghawar field. With 260 billion barrels of reserves, the first year of decline would be 2043 for the 10 million b/d case and 2032 for the 12 million b/d case.[2] Production at 15 million b/d could not be sustained for more than a year or two. Figure A3-2 compares these production profiles with domestic Saudi consumption.

1. Nasser and Saleri, 2004 (Saudi Aramco presentation).

2. These correspond to the point in the Saudi Aramco presentation (Nasser and Saleri, 2004) at which additional reserves would be needed to sustain production.

Figure A3-2. Saudi Arabia: Oil Production Profiles and Consumption, 2000–60

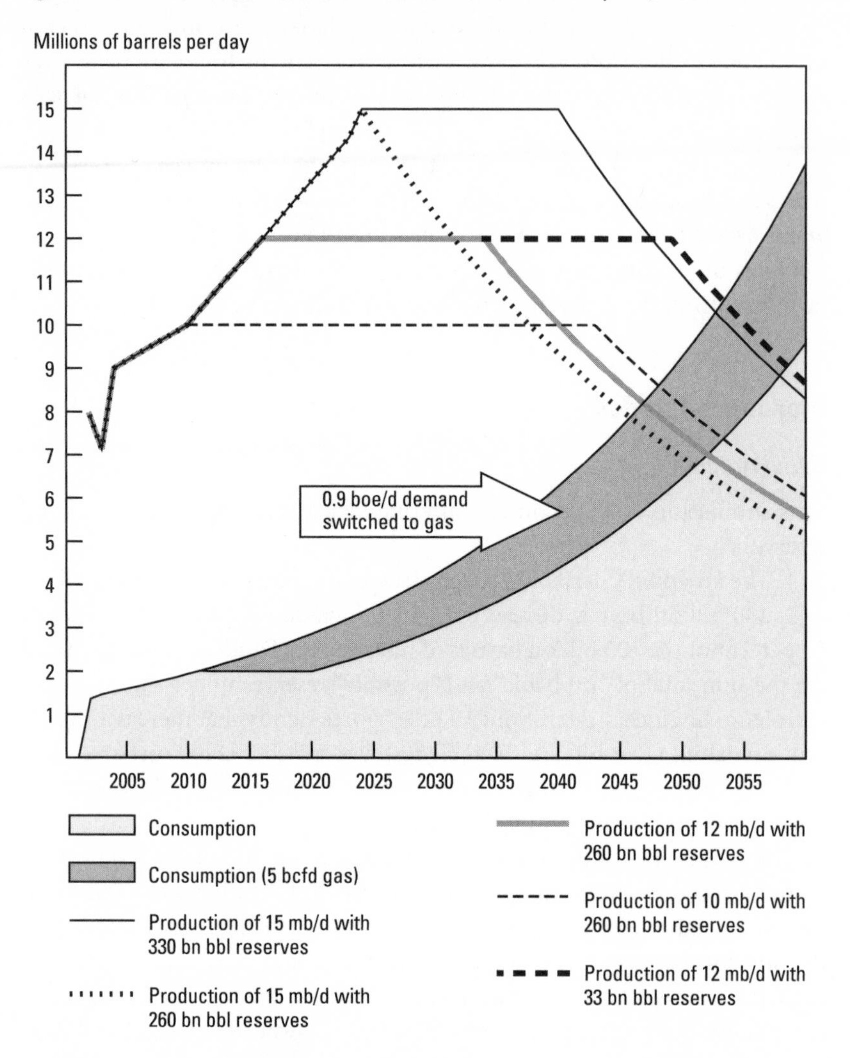

It is accepted that the country's domestic energy consumption will track non-oil GDP at its assumed growth rate of four percent per annum. But this is conservative: at a high level of per capita income such as Saudi Arabia's, energy demand usually grows faster than GDP, and at present this demand is supplied almost entirely by oil. The simulation also assumes that between 2010 and 2020, 900,000 b/d of oil demand is substituted by 880,000 barrels of oil equivalent (boe) of natural gas, in line with Saudi Aramco's objectives.

The natural gas share of the Saudi energy market would then be just under 25 percent. After 2020, this share is held constant and oil demand grows with non-oil GDP.

The dates of inflection in the production profiles are the outcomes of the assumptions of the model. For example, it is assumed that the gas strategy allows the substitution of gas for oil in domestic consumption in order to hold domestic oil demand flat during the period 2012–20.

Iran

In 2004, Iran's oil reserves were 132 billion barrels[3] and its gas reserves were 236.7 trillion cubic meters.[4] Its natural gas liquid reserves were 4.3 trillion boe in 2000.[5]

Its oil production is assumed to increase from 3.7 million b/d in 2003 to 7 million b/d by 2024.[6] This rate can be sustained, if there are no further additions to reserves, until 2030 with a depletion rate of just over 3 percent (an R/P of 30:1). Beyond 2030, 7 million b/d can be sustained to 2040 only if the R/P is allowed to fall to 10:1 (typical of North America and Europe) and if it is technically possible to do so.[7] Thereafter, oil production will inevitably decline. (For a profile of Iran's consumption and export of oil and gas to 2030, see figure A3-3.)

Gas production in Iran is projected to increase from 79 billion cubic meters in 2003 to 97 billion cubic meters in 2009 and 360 billion cubic meters in 2020. This increase includes LNG and gas-to-liquid (GTL) conversion but excludes re-injection, which, from 2010, is assumed to equal the volume of gas produced for the market.[8] Ultimately, a proportion of the gas re-injected before 2030 will be produced after 2030.

NGL production is assumed to be constant at 300,000 boe/d until 2033, after which it declines at 10 percent per annum, to maintain an R/P ratio of 10:1.

3. 2004: Iranian Petroleum Ministry announcement, *EIA Country Analysis Brief—Iran*, August 2004.

4. 2000–03: *BP Statistical Review*, 2004.

5. "Known NGLs" median estimate, U.S. Geological Survey, 2000.

6. Quoted in the *EIA Country Analysis Brief—Iran*, August 2004.

7. The known gas reserves may not be able to sustain increased gas re-injection and support 7 million b/d.

8. The projection to 2009 is based on a presentation by H. Kazempour Ardebili to the Oxford Energy Seminar in September 2004. The projection to 2020 is derived from a presentation by Rokneddin Javadi, Managing Director of the National Iranian Gas Export Company, to the Iranian Export Gas Conference, April 2004, as reported in *Gas Matters*, May 2004.

Figure A3-3. Iran's Oil and Gas: A Profile of Their Consumption and Export, 2000–30

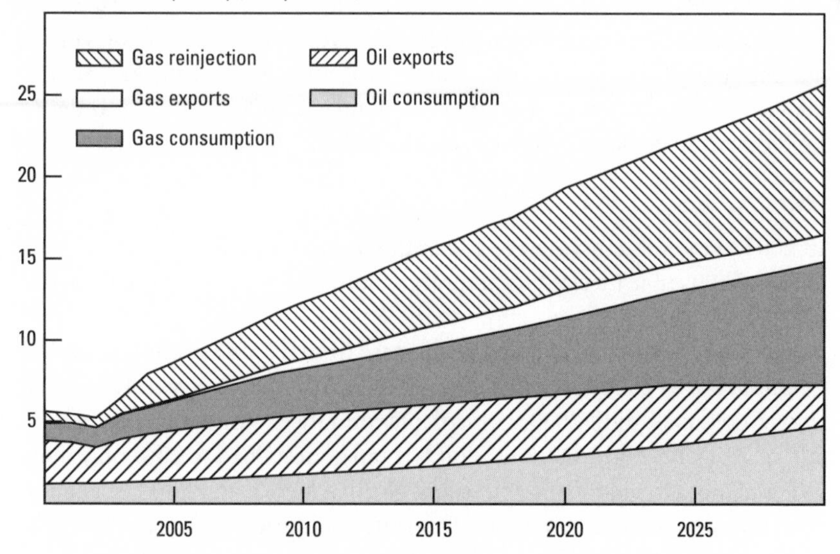

Gas consumption in Iran is assumed to increase more rapidly than oil consumption until 2009, expanding its share of the domestic energy market from about 50 percent to just over 60 percent by 2010.[9] Thereafter, the model assumes that both gas and oil consumption increase in line with the non-oil GDP at 5 percent per annum.

Algeria

Algeria's oil reserves in 2004 were 1.4 million tons (11.3 billion barrels),[10] while its gas reserves were 4.5 trillion cubic meters (160 trillion cubic feet).[11]

The country's oil production is assumed to grow until the R/P ratio falls to 10:1 (75 million tons per annum) in 2006. This assumes a rate of depletion similar to that in the U.S. and Europe, and much more aggressive than that assumed for the Middle East producers. After 2006, oil production is reduced every year in order to maintain that ratio. However, with liquids (LPG and NGLs) contributed from increasing gas production, total liquids production

9. This assumption is based on Ardebili's figures for gas and the model assumptions for oil.

10. In 2000, the U.S. Geological Survey assessed probable undiscovered liquids (crude and natural gas liquids) in Algeria as 1.2 billion tons (8.8 billion barrels).

11. The U.S. Geological Survey assessed its mean undiscovered gas to be 1.4 trillion cubic meters (49.3 trillion cubic feet).

of just under 100 million tons per annum would be sustained until 2012. A plateau for gross gas production is set, corresponding to around 130 billion cubic meters of marketable gas,[12] which would be reached in 2015 and sustained until 2026. Algerian consumption is given first call on production, so that exports reflect the full effect of declining supply as production comes off its plateau. Oil exports decline from 2006, gas exports from 2016.

Figure A3-4 shows a "worst-case" profile, because it assumes that no new reserves are found. A 10 percent increase in reserves would delay the points of inflection by about two years, depending on when the reserves were discovered and how long it took to bring them into production.

Kuwait

Kuwait's oil reserves in 2004 were 99 billion barrels (including its share of the Neutral Zone).

The production simulation in figure A3-5 is simple. It uses the stated KPC objective of increasing oil production capacity to 4 million b/d by 2020 and assumes that this capacity is fully utilized from 2005 onward. With no additions, this plateau could be maintained at a 3 percent depletion rate until about 2040. With depletion rates approaching 10 percent, as in mature producing countries, the plateau could be maintained even longer. After the plateau of production is reached in 2020, rising consumption would lead to falling exports, but the level of consumption would be low. Only a small amount of gas is produced in Kuwait. As this is likely to remain small, gas is ignored.

The United Arab Emirates

There are two production scenarios for the UAE:

1. The "low case." Oil production is increased to 3 million b/d by 2010, with 600,000 b/d of condensates. Annual gas production, net of reinjection, increases to 74 billion cubic meters by 2010. It then increases in line with annual consumption (at 5 percent) while gas for exports is held constant at 40 billion cubic meters (slightly above current levels), rising to 130 billion cubic meters per annum by 2030. This rate of oil production could be maintained at a 5 percent depletion rate (20:1 R/P) well into the second half of the century, and the rate of gas production at least until 2050. As there is no doubt that there will be additions to both oil and gas reserves, this represents a worst-case production scenario.

12. This figure is net of re-injection, shrinkage, losses and energy costs in liquefaction.

Figure A3-4. Algeria's Hydrocarbons: Production and Consumption, 2000–30

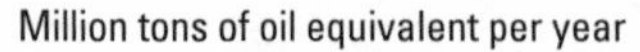

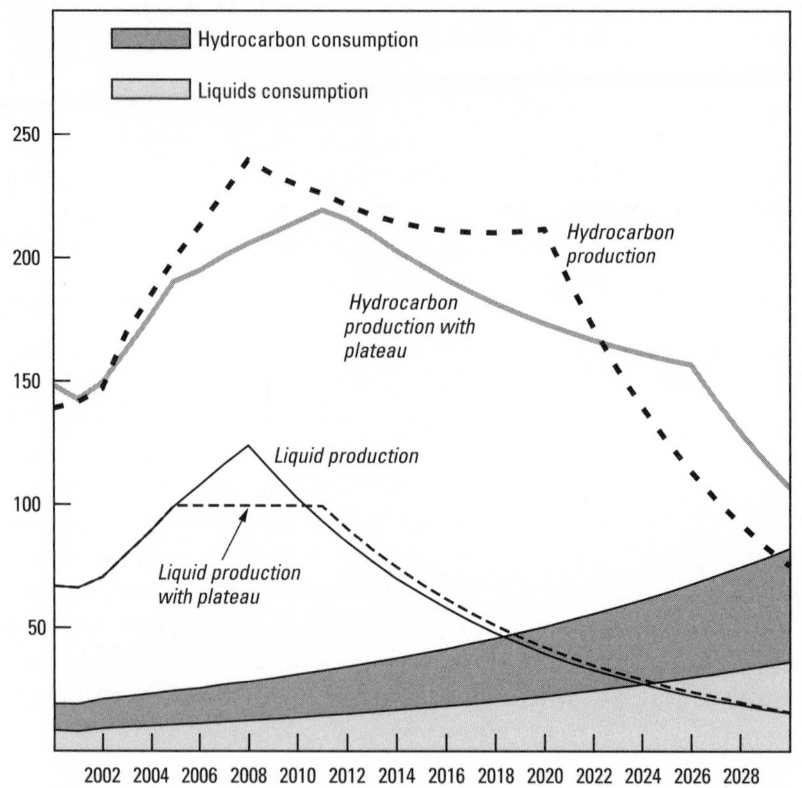

2. The "high case." With the same reserves limits, oil production increases to 5 million b/d by 2030, with condensate production stable at 1.3 million b/d. Assuming reserves of 97 billion barrels (as currently known), this rate of production could be maintained until 2043, after which, in order to keep the depletion rate at 5 percent (for a 20-year reserve), production would decline. Gas production is taken to 140 billion cubic meters by 2030, and exports would increase to around 55 billion cubic meters in the mid-2020s. With currently known gas reserves, this rate of production could not be maintained past the mid-2030s.

Figures A3-6 and A3-7 show the production profiles to 2030 for the Emirates' oil and condensates and their gas and NGLs respectively.

Figure A3-5. Kuwaiti Oil: Production, Consumption, and Exports, 2000–45

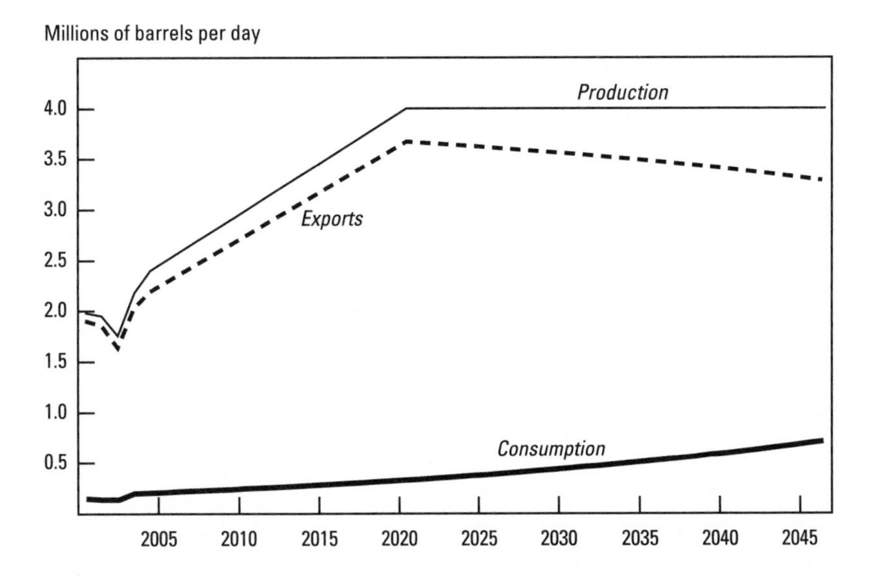

Figure A3-6. United Arab Emirates: Production and Consumption of Oil and Condensates, 2000–30

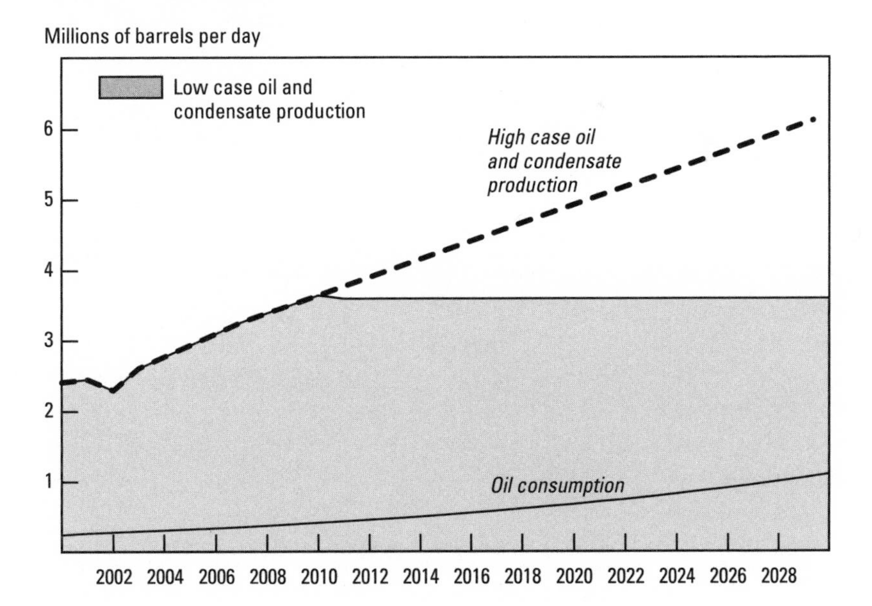

Figure A3-7. United Arab Emirates: Production and Consumption of Gas and NGLs, 2000–30

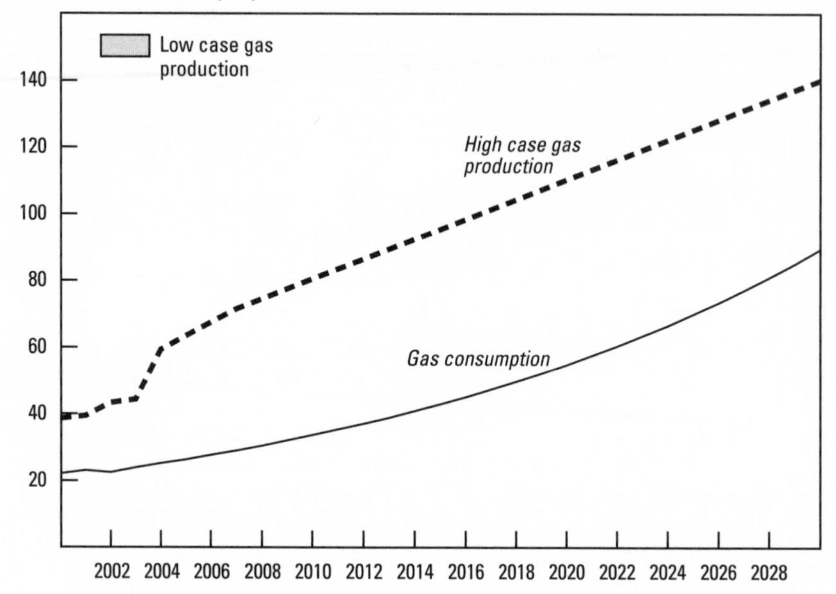

Bibliography

Books, Papers, and Reports

Abbasi, M. J., Mehryar, A., Jones, G. and McDonald, P. (2001): "Revolution, War and Modernization: Population Policy and Fertility Change in Iran," paper prepared for the Twenty-Fourth IUSSP General Conference, Salvador, Brazil, 18–24 August.

Adelman, M. A. (1972): *The World Petroleum Industry*, Baltimore, MD: Johns Hopkins University Press.

ADNOC (2005): *ADNOC's Five-Year Achievements Report: 2000–2004*, Abu Dhabi.

Aïssaoui, Ali (2005): "Financing Saudi Oil and Gas-Related Projects: Issues and Trends," presentation given at the MEED Arab Petroleum Investments Corporation Conference, "Major New Project Opportunities in the Eastern Province," Bahrain, 1–2 March.

———. (2001): *Algeria: The Political Economy of Oil and Gas*, Oxford: Oxford Institute for Energy Studies.

———. (2000): "Managing Hydrocarbon Resources in a New Era: The Call from Algeria," *Oxford Energy Comment*, Oxford Institute for Energy Studies, July. Available at *http://www.oxfordenergy.org/comment.php?0007*.

Anastassopoulos, Jean-Pierre, Blanc, Georges and Dussauge, Pierre (1987): *State-Owned Multinationals*, Chichester: John Wiley & Sons.

Arad, R.W. et al. (1979): *Sharing Global Resources*, New York: McGraw-Hill.

Arnott, Robert (2004): "National Oil Company Websites as Primary Sources of Oil and Gas Information," *Journal of Energy Literature*, 10 (1): 22–36.

Bamberg, J. H. (1994): *The History of the British Petroleum Company: Volume 2, The Anglo-Iranian Years, 1928–1954*, Cambridge: Cambridge University Press.

Banque d'Algérie (2004): *Indicateurs monétaires et financiers*, 2ième trimestre.

Barnett, Steven A. and Ossowski, Rolando J. (2002): "Operational Aspects of Fiscal Policy in Oil-Producing Countries," *IMF Working Paper No. 02/177*, Washington, D.C.

Bill, James A. and Louis, William Roger (eds.) (1988): *Musaddiq, Iranian Nationalism, and Oil*, London: I.B. Tauris.

Bouhafs, Abdelhak (former CEO of Sonatrach) (1994): "Stratégie de modernisation de Sonatrach," *Économies et Sociétés*, No. 6, Série EN, September.

Boussena, Sadek (former Minister of Energy of Algeria) (1994): "L'adaptation des companies nationales au nouveau contexte pétrolier," *Économies et Sociétés*, No. 6, Série EN, September.

Brand, Laurie A. (1994): *Jordan's Inter-Arab Relations: The Political Economy of Alliance Making*, New York: Columbia University Press.

British Embassy, Abu Dhabi & Trade Partners UK (2003): *Oil, Gas and Petrochemicals Sector Report: Abu Dhabi*. Available at *http://www.agcc.co.uk/documents/Oil_Gas_&_Petrochem_Sector_Report_Abu_Dhabi_BE_August_2003.doc.*

BP (2004, 2005): *BP Statistical Review of World Energy*.

Brynen, Rex (1992): "Economic Crisis and Post-Rentier Democratization in the Arab World: The Case of Jordan," *Canadian Journal of Political Science*, 25 (1), March.

Central Bank of Kuwait (2003), *Annual Report 2003*.

———. (2004), *Quarterly Statistical Bulletin*, December.

Central Bank of Iran (2004): *Economic Trends 35*, Fourth Quarter, 1382.

Charles Rivers Associates (2002): "The Next Generation National Oil Company," November.

Chaudhry, Kiren Aziz (1997): *The Price of Wealth: Economies and Institutions in the Middle East*, Ithaca, NY: Cornell University Press.

Chevalier, Jean-Marie (1994): "L'avenir des sociétés nationales des pays exportateurs d'hydrocarbures," *Économies et Sociétés*, No. 6, Série EN, September.

Clark, John G. (1990): *The Political Economy of World Energy: A Twentieth Century Perspective*, New York: Harvester Wheatsheaf.

Clawson, Patrick (2004): "The Paradox of Anti-Americanism in Iran," *Middle East Review of International Affairs*, 8 (1), March.

Cordesman, Anthony H. and Burke, Arleigh A. (2002): *Saudi Arabia Enters the 21st Century*, Center for Strategic and International Studies, Washington, D.C., 30 October.

Crystal, Jill (1990): *Oil and Politics in the Gulf: Rulers and Merchants in Kuwait and Qatar*, Cambridge: Cambridge University Press.

Davis, J. M., Ossowski, R. and Fedilino, A. (2003): *Fiscal Policy Formulation and Implementation in Oil-Producing Countries*, International Monetary Fund.

Dazi-Heni, Fatiha (2001): "Le pouvoir Al-Saoud face aux enjeux d'une société urbaine en mutation," *Maghreb-Machrek* (numéro spécial : l'Arabie saoudite et la péninsule après le 11 septembre ; défis et enjeux d'une région en crise), 174: 8–17.

Dialdin, Ali M. and Tahlawi, Muhammad A. (eds.) (1998): *Saudi Aramco and Its People: A History of Training*, Dhahran: The Saudi Arabian Oil Company.

Diwan, Ishac and Girgis, Maurice (2002): "Labor force and development in Saudi Arabia," paper presented at *Symposium on the Future Vision for the Saudi Economy*, Riyadh, 19–23 October.

Economist Intelligence Unit (2004): *Iran Country Report 2004–2005*; *Algeria Country Report 2004–2005*; *Saudi Arabia Country Report 2004–2005*, London: Economist Group Limited.

Energy Information Administration (EIA), Department of Energy, United States Government, *Country Analysis Briefs* (various years). Available at *www.eia.doe.gov*.

European Union (2003): "Accession of Saudi Arabia to the WTO," European Union at the United Nations, 30 August. Available at *http://europa-eu-un.org/articles/lt/article_2667_lt.htm.*

Farjadi, Gholamali and Pirzadeh, Ali (2003): "Labor Market and Employment in Iran," Economic Research Forum Paper. Available at *http://www.erf.org.eg/grp/GRP_Sep03/Iran-Labor.pdf*.

Farmanfarmaian, Manucher (1997): *Blood and Oil: Memoirs of a Persian Prince*, New York: Random House.

Fromkin, David (2nd ed., 2003): *A Peace to End All Peace; The Fall of the Ottoman Empire and the Creation of the Modern Middle East*, London: Phoenix.

Gardner, Edward (2003): "Creating Employment in the Middle East and North Africa," Washington, D.C.: International Monetary Fund.

Gause III, F. Gregory (1994): *Oil Monarchies: Domestic and Security Challenges in the Arab Gulf States*, New York: Council on Foreign Relations Press.

Grayson, Lesley E. (1981): *National Oil Companies*, Chichester: John Wiley and Sons.

Hamilton, Kirk (2002): "Accounting for Sustainability," Washington D.C.: World Bank, July. Available at *http://www.chinaeol.net/wbi/eep/docs/reading/Accounting%20For%20Sustainbility.pdf.*

Higgins, Gordy (1999–2000): "A Review of Privatization Definitions, Options, and Capabilities for the Business, Labor and Agriculture Interim Committee," Washington, D.C.: World Bank. Available at *http://leg.state.mt.us/content/publications/research/past_interim/defined.pdf*

Hollis, Rosemary (ed.) (1998): *Oil and Regional Developments in the Gulf*, London: Royal Institute of International Affairs.

Howarth, Stephen (1997): *A Century in Oil: The Shell Transport and Trading Company 1897–1997*, London: Weidenfeld & Nicolson.

International Energy Agency (1995): *Middle East Oil and Gas 1995*, Paris: OECD/IEA.

———. (2004): *World Energy Outlook*, Paris: OECD/IEA.

International Monetary Fund (IMF) (2000): "IMF Concludes Article IV Consultation with Kuwait," *Public Information Notice 00/27*, Washington D.C. Available at *http://www.imf.org/external/np/sec/pn/2000/pn0027.htm.*

———. (2003a): "IMF Concludes 2003 Article IV Consultation with Saudi Arabia," *Public Information Notice 03/143*, Washington D.C. Available at *http://www.imf.org/external/np/sec/pn/2003/pn03143.htm.*

———. (2003b): "IMF Concludes 2003 Article IV Consultation with Algeria," *Public Information Notice 04/03*, Washington D.C. Available at *http://www.imf.org/external/np/sec/pn/2004/pn0403.htm.*

———. (2004a): "Islamic Republic of Iran Statistical Appendix," *Country Report No. 04/307*, Washington, D.C. Available at *http://www.imf.org/external/pubs/ft/scr/2004/cr04307.pdf.*

———. (2004b): "Islamic Republic of Iran: Selected Issues," *Country Report No. 04/308*, Washington, D.C. Available at *http://www.imf.org/external/pubs/ft/scr/2004/cr04308.pdf.*

———. (2004c): "IMF Concludes 2004 Article IV Consultation with the Islamic Republic of Iran," *Public Information Notice 04/109*, Washington, D.C. Available at *http://www.imf.org/external/np/sec/pn/2004/pn04109.htm.*

———. (2004d): "Kuwait: Statistical Appendix," *Country Report No. 04/197*, Washington, D.C. Available at *http://www.imf.org/external/pubs/ft/scr/2004/cr04197.pdf.*

———. (2004e): "United Arab Emirates Statistical Appendix," *Country Report No. 04/174*, Washington, D.C. Available at *http://www.imf.org/external/pubs/ft/scr/2004/cr04174.pdf.*

———. (2005a): "Algeria: 2004 Article IV Consultation—Staff Report," *Country Report No. 05/50*, Washington, D.C. Available at *http://www.imf.org/external/pubs/ft/scr/2005/cr0550.pdf.*

———. (2005b): "Algeria Statistical Appendix," *Country Report No. 05/51*, Washington, D.C. Available at *http://www.imf.org/external/pubs/ft/scr/2005/cr0551.pdf.*

———. (2005c): "Kuwait Statistical Appendix," *Country Report No. 05/234*, Washington, D.C. *Available at http://www.imf.org/external/pubs/ft/scr/2005/cr05234.pdf.*

Khavand, Fereydoun (2004): "Soviet Era Planning: The Plight of the Fourth Five-Year Plan," Radiofarda, U.S. International Broadcasting, 21 June. Available at *http://www.radiofarda.com/transcripts/topstory/2004/06/20040621_1430_0421_0750_EN.asp.*

Khelil, Chakib (Minister of Energy and Mines and then CEO of Sonatrach) (2002): "Welcoming remarks at the National Oil Companies Forum," Algiers, April 26–27.

Kissinger, Henry (1994): *Diplomacy*, New York: Touchstone.

Kuwait Petroleum Corporation (2003): *Annual Report 2001–2002.*

———. (no date): *Kuwait Petroleum Corporation*, Media Department Information and Documentation.

Luciani, Giacomo (1990): "Arabie Saoudite: l'industrialisation d'un État allocataire," *Maghreb-Machrek*, Vol. 129, September.

———. (1984): *The Oil Companies and the Arab World*, London: Croom Helm.

Mabro, Robert, Sandvold, Tore I., Manso, Rogerio, Brandao, Fabio and Luciani, Giacomo (2004): "NOC to IOC? National Oil Companies and International Oil Companies," *Oxford Energy Forum*, Issue 57, May.

———. (2003): "Setting the Scene," *OPEC Review*, Vol. 27, No. 3, Special Issue—Workshop Proceedings, Joint OPEC/IEA Workshop on Oil Investment Prospects, 25 June.

Madelin, Henri (1975, English trans.): *Oil and Politics*, Farnborough, U.K.: Saxon House.

Marcel, Valérie (2004): "National and International Oil Companies: Existing and Emerging Partnerships," paper given at the Emirates Centre for Strategic Studies and Research's Tenth Energy Conference in Abu Dhabi, 26–27 September (ECSSR publication forthcoming).

———. (2005): "Good Governance of the National Oil Company," Chatham House, March. Available at *http://www.chathamhouse.org.uk/pdf/research/sdp/GGjmc.doc.*

———. (2005): "The internationalization of NOCs," *First Magazine*, 16–17 October.

Marschall, Christin (2003): *Iran's Persian Gulf Policy: From Khomeini to Khatami*, New York: RoutledgeCurzon.

Ministry of Energy and Mines, People's Democratic Republic of Algeria (2005): "Directive relative à la promotion de l'emploi féminin," No. 01 CAB/SB, Algiers, 5 January.

Ministry of Finance, People's Democratic Republic of Algeria (2004): *Direction Générale des Études et de la Prévision de la Situation Économique et Financière en 2003*, Algiers.

Ministry of Information and Culture, United Arab Emirates (2005): *United Arab Emirates Yearbook 2005*. Available at *http://www.uaeinteract.com/uaeint_ misc/pdf_ 2005/index.asp#year.*

Ministry of Petroleum, Islamic Republic of Iran (2001–02, 2002–03): *Financial Statement.*

———. (1997–98, 1998–99, 2000–01): *Iranian Petroleum Industry Annual Report.*

Mitchell, John V. (2004): *Petroleum Reserves in Question*, Briefing Paper SDP 04/03, Chatham House, October.

———. (2005): "Background Paper—Section 3: Demand and Market Conditions" (draft), paper prepared for "International Energy Dialogue," Chatham House–IEF Workshop, London, 27–28 April.

Moghaddam, Mohammad Reza (2003): "Improving Iran's Domestic Energy Basket," Ph.D. dissertation, Tilburg University.

Mommer, Bernard (2002): *Global Oil and the Nation State*, Oxford: Oxford University Press.

Morse, Edward L. (1994): "State Control over the Energy Sector: The Ambiguous Roles of National Oil Companies," *Économies et Sociétés*, No. 6, Série EN, September.

Myers, Keith and Carpentier, Philippe (2004): "Success in European Market Joint Ventures," *Wood Mackenzie Horizons*, Energy Issue 19.

Naimi, Ali (Minister of Petroleum and Mineral Resources, Kingdom of Saudi Arabia) (2004): Keynote address delivered at "Oil, Economic Change and the Business Sector in the Middle East," Chatham House conference, London, 29–30 November.

Al-Nasr, Nadhmi A. (Executive Director of Community Buildings & Office Services, Saudi Aramco) (2004): "Saudi Aramco and the Domestic Private Sector," paper presented at "Oil, Economic Change and the Business Sector in the Middle East," Chatham House conference, London, 29–30 November.

Nasser, Amin H. and Saleri, Nansen G. (Saudi Aramco) (2004): Paper presented at Fifth International Energy Summit, Paris, 29 April.

Olorunfemi, Michael A. (1991): "The Dynamics of National Oil Companies," *OPEC Review*, 15 (4), 321–33.

Organization for Economic Cooperation and Development (OECD) and African Development Bank (2004): "African Economic Outlook 2003/2004—22 Country Studies."

Organization of the Petroleum Exporting Countries (OPEC) (2003, 2004): *OPEC Annual Statistical Bulletin*, Vienna.

———. (2004): *Who Gets What from Imported Oil*, Vienna. Available at *http://www.opec.org/library/Special%20Publications/pdf/2004.pdf.*

———. (undated): *General Information and Chronology: 1960–1992*, Vienna.

Pakkiasamy, Divya (2004): *Saudi Arabia's Plan for Changing Its Workforce*, Migration Information Source, Migration Policy Institute, Washington, D.C., 1 November.

Petrochemicals Industries Company (PIC) (2003): *Annual Report 2002–2003.*

Philip, George (1994): *The Political Economy of International Oil*, Edinburgh: Edinburgh University Press.

Al-Sabah, Y. S. F. (1980): *The Oil Economy of Kuwait*, London: Kegan Paul.

Saudi American Bank (2005): *The Saudi Economy 2004 Performance, 2005 Forecast*, February.

Saudi Arabian Monetary Agency (2003a): "The Implications of the New Capital Market Law on the Future of Investment Opportunities," paper presented at SAGIA Annual Retreat, The Progress of Privatization, Rome, 14–15 October. Available at *http://www.sama.gov.sa/en/news/2003-10/2003-10-14.htm.*

———. (2003b): *SAMA 39th Annual Report 1423/24H.*

Saudi Aramco. (2003, 2004): *Saudi Aramco Facts and Figures 2002, 2003.* Available at *www.saudiaramco.com/bvsm/JSP/content/channelDetail.jsp?SA.channelID=-11710-34k.*

Simmons, Matthew R. (2004): "The Saudi Arabian Oil Miracle," presentation given at the Center for Strategic and International Studies, Washington D.C., 24 February.

Sonatrach (2001, 2002, 2003, 2004): *Rapport Annuel & Rapport Financier.*

Stern, Jonathan (2002): *Security of European Natural Gas Supplies: The Impact of Import Dependence and Liberalization*, Royal Institute of International Affairs, July.

Stevens, Paul (2003): "National Oil Companies: Good or Bad? A Literature Survey" (draft), World Bank workshop on "National Oil Companies: Current Roles and Future Prospects," Washington, D.C., 27 May.

Subroto, Roberto (former Secretary-General of OPEC) (1994): "The Future of National Oil Companies of OPEC," *Économies et Sociétés*, No. 6, Série EN, September.

Tétreault, Mary-Ann (1995): *The Kuwait Petroleum Corporation and the Economics of the New World Order*, Westport, CT: Quorum Books.

———. (2000): *Stories of Democracy: Politics and Society in Contemporary Kuwait*, New York: Columbia University Press.

———. (2003): "Pleasant Dreams: The WTO as Kuwait's Holy Grail," *Critical Middle Eastern Studies*, 12 (1), Spring.

———. (2005): "Women's Rights and the Meaning of Citizenship in Kuwait," Tharwa Project, 10 February. Available at *http://www.tharwaproject.com/index.php?option=com_keywords&task=view&id=1299&Itemid=0.*

Total (2003): *Key Figures 2003.*

United Nations Conference on Trade and Development (UNCTAD) (2004): "Natural Gas Reserves," Market Information in the Commodities Area. Available at *http://r0.unctad.org/infocomm/anglais/gas/market.htm.*

United Nations Development Program (UNDP) (2003, 2004, 2005): *Arab Human Development Report.*

U.S. and Foreign Commercial Service and U.S. Department of State (2003): "Kuwait Country Commercial Guide FY 2004: Marketing U.S. Products & Services," 2 September.

Williamson, Edwin (1992): *The Penguin History of Latin America*, London: Penguin Books Ltd.

Williamson, John (2000): "What Should the World Bank Think of the Washington Consensus?" *World Bank Observer*, 15, 2 August.

World Bank (2003a): "People's Democratic Republic of Algeria: A Medium-Term Macroeconomic Strategy for Algeria," *World Bank Report 26005-AL.*

———. (2003b): "Iran: Medium Term Framework for Transition: Converting Oil Wealth to Development," *World Bank Report 25848-IRN.*

———. (2004): *To 2015: World Development Indicators 2004*, Washington D.C.

———. (2005a): *World Development Indicators 2005*, Washington D.C.

———. (2005b): "The Impact of Higher Oil Prices on Low Income Countries and on the Poor," UNDP/ESMAP (United Nations Development Program/World Bank Energy Sector Management Assistance Program), Washington, D.C, March. Available at *http://wbln0018.worldbank.org/esmap/site.nsf/files/299-05_HigherOilPrices_ Bacon.pdf /$FILE/299-05_HigherOilPrices_Bacon.pdf.*

Yamani, Mai (2000): *Changed Identities: The Challenge of the New Generation in Saudi Arabia*, London: Royal Institute of International Affairs.

Yergin, Daniel (1991): *The Prize: The Epic Quest for Oil, Money, and Power*, New York: Simon & Schuster.

Zanoyan, Vahan (2000): "The Changing Role of Oil," paper presented at "Pioneering Mideast Economic Transformation: Three Views," U.S.–GCC Corporate Cooperation Committee and its Secretariat, National Council on U.S.–Arab Relations, 25 September.

———. (2004): "Institutional Cooperation in Oil and Gas: Governments, Companies and the Investment Climate," Ninth International Energy Forum, Amsterdam, 22–23 May.

Periodicals, Special Series, and Papers

Arab Oil and Gas Directory
The Economist
Energy Compass
International Petroleum Encyclopaedia 2004
Middle East Economic Digest (*MEED*)
Middle East Economic Survey (*MEES*)
Oil and Gas Abacus
Oil and Gas Journal
Petroleum Intelligence Weekly

Literature Survey

Sources in French

Blin, Louis (1996): *Le pétrole du Golfe; Guerre et paix au Moyen-Orient*, Paris: Maisonneuve et Larose.

Blin, Louis et Fargues, Philippe (eds.) (1995): *L'économie de la paix au Proche-Orient*, Paris: Maisonneuve et Larose.

Bonnenfant, Paul (ed.) (1982): *La péninsule Arabique d'aujourd'hui*, tomes I–II, Paris: Éditions du CNRS.

Bourgey, André (ed.) (1982): *Industrialisation et changements sociaux dans l'Orient arabe*, Beirut: CERMOC.

Chatelus, Michel (1974): *Stratégie pour le Moyen-Orient*, Paris: Calmann-Lévy.

———. (2001): "La situation économique des pays producteurs de pétrole de la péninsule arabique," *Maghreb-Machrek* (numéro spécial : l'Arabie saoudite et la péninsule après le 11 septembre ; défis et enjeux d'une région en crise), n° 174, 58–64.

Chevalier, Jean-Marie (1973): *Le nouvel enjeu pétrolier*, Paris: Calmann-Lévy.

Corm, Georges (1999): *Le Proche-Orient éclaté*, Paris: Gallimard.

Dazi-Heni, Fatiha (2001): "Le pouvoir Al-Saoud face aux enjeux d'une société urbaine en mutation," *Maghreb-Machrek* (numéro spécial : l'Arabie saoudite et la péninsule après le 11 septembre ; défis et enjeux d'une région en crise), n° 174, 8–17.

Destremau, Blandine (dir.) (2000): "Formes et mutations des économies rentières au Moyen-Orient," *Revue Tiers Monde* (numéro spécial), n° 163, July–September.

Économies et Sociétés (1994): "L'avenir des sociétés nationales des pays exportateurs d'hydrocarbures" (numéro spécial), n° 6, Série EN, September.

Fargues, Philippe (1980): *Réserves de main d'œuvre et rente pétrolière: étude démographique des migrations de travail vers les pays arabes du Golfe*, Beirut: CERMOC.

———. (1983): *L'aide des pays de l'OPEP*, Paris: OECD.

———. (2000): *Générations arabes: L'alchimie du nombre*, Paris: Fayard.

Khader, Bichara (ed.) (1981): *Monde arabe et développement économique*, Paris: Le Sycomore.

Maachou, Abdelkader (1982): *l'OPEP et le pétrole arabe*, Paris: Berger-Levrault.

Madelin, Henri (1973): *Pétrole et politique en Méditerranée occidentale*, Paris: Arman Colin.

Mahiou, Ahmed et Henry, Jean-Robert (eds.) (2001): *Où va l'Algérie?*, Paris: Karthala-IREMAM.

Nouschi, André (1972): *Luttes pétrolières au Proche-Orient*, Paris: Flammarion.

———. (1999): *Pétrole et relations internationales depuis 1945*, Paris: Arman Colin.

Sader, Makram (1983): *Le développement industriel de l'Irak*, Beirut: CERMOC.

Sarkis, Nicolas et al. (1968): *Pétrole et développement économique au Moyen-Orient*, Paris: Mouton.

Sid Ahmed, Abdelkader (1980): *L'O.P.E.P. Passé, présent et perspectives (Éléments pour une économie politique des économies rentières)*, Paris: Economica.

———. (1983): *Développement sans croissance ; l'expérience des économies pétrolières du tiers monde*, Paris: Publisud.

Sources in Arabic

Abdallah, Hussein (2000): *The Future of Arab Oil*, Beirut: Center for Arab Unity Studies.

Abdel-Fadil (6th ed., 2000): *Oil and Arab Unity: The Impact of Arab Oil on Future Arab Unity and Economic Relations*, Beirut: Center for Arab Unity Studies.

Bouheiry, Marwan (2nd ed., 1986): *Arab Oil and the Threats of American Intervention, 1973–1979*, Beirut: Institut d'études palestiniennes.

Al-Cheikh, Toufiq (1988): *Oil and Politics in the Kingdom of Saudi Arabia*, London: Dar Al-Safa.

Corm, Georges (1977): *A Challenge to the Arab Economy: Studies of Petroleum Economics, Finance and Technology*, Beirut: Dar Al-Tali'at.

———. (1979): *Arab Oil and the Palestinian Question*, Beirut: Institut d'études palestiniennes.

Al-Fares, Abdel-Rasaq (1996): *Wasted Energy: The Development and the Problem of Energy in the Arab Nation*, Beirut: Center for Arab Unity Studies.
Al-Kouari, Hussein (2nd ed., 1996): *The Lost Development! Or the Lost Opportunities for Development*, Beirut: Center for Arab Unity Studies.
Al-Qadi, Loubna (1985): *Rapid Development in Some Oil Producing Countries of the Arab Gulf*, Kuwait: Kuwait Institute for Scientific Progress.
Al-Ramihi, Mohamed (1995a): *The Gulf Is Not Oil; Study of the Problem and the Development of Unity*, Beirut: Dar Al-Jadid.
———. (1995b): *Oil and Social Change in the Arab Gulf*, Beirut: Dar Al-Jadid.
Rida, Hilal (1991): *The Conflict over Kuwait: A Question of Security and Riches*, Beirut: Dar Al-Sinai Editions.
Saaddine, Ibrahim (ed.) (1989): *Arab Development*, Beirut: Center for Arab Unity Studies.
Al-Sabi'i, Abdallah (1989) (2nd ed.): *The Discovery of Oil and Its Effect on the Economic Life of the Oriental Region, 1933–1960*, Kingdom of Saudi Arabia, Dar Al-Sharif.
Sayigh, Yusif (1986) (2nd ed.): *Arab Oil and the Palestinian Question in the 1980s*, Beirut: Institut d'études palestiniennes.
———. (1992): *Untamed Development: From Dependency to the Capacity to Count on One's Own Strength in the Arab World*, Beirut: Center for Arab Unity Studies.
———. (1996): *Arab Development: From Past Mistakes to Fear of the Future*, Amman: Mountada Al-Fikr Al-'Arabi.

Official Hydrocarbon Sector Websites

Algeria, Ministry of Energy and Mines: *www.mem-algeria.org*
Abu Dhabi National Oil Company: *www.adnoc.com*
Islamic Republic of Iran, Ministry of Petroleum: *www.nioc.org*
Kingdom of Saudi Arabia, Ministry of Petroleum and Mineral Resources: *www.mopm.gov.sa*
Kuwait Petroleum Corporation: *www.kpc.com.kw*
National Iranian Oil Company: *www.nioc.com*
Saudi Aramco: *www.saudiaramco.com*
Sonatrach: *www.sonatrach-dz.com*

About the Authors

VALÉRIE MARCEL is Principal Researcher with the Energy, Environment and Development Programme at Chatham House, where she has led energy research since 2002 and published on the politics of oil in the Middle East. She has carried out extensive fieldwork in the major oil-producing countries of the region. She is also leading a major initiative on "Good Governance of the National Petroleum Sector," with the aim of working closely with the governments of oil- and gas-producing countries, their national oil companies and other stakeholders to establish principles of good governance and guidelines for best practice. Her publications on this subject include *Good Governance of the National Oil Company* (Chatham House Report, March 2005). Dr. Marcel's work at Chatham House previously focused on Iraq's oil industry and the political issues affecting future prospects for the industry, with publications including *Total in Iraq* (The Brookings Institution, August 2003); *Iraq's Oil Tomorrow* (RIIA Report, coauthored with John Mitchell, April 2003); and *The Future of Oil in Iraq* (RIIA Briefing Paper, December 2002). She was previously Lecturer in International Relations at the Institut d'études politiques de Paris and the University of Cairo.

JOHN V. MITCHELL is an Associate Fellow of the Energy, Environment and Development Programme at Chatham House and Research Adviser at the Oxford Institute of Energy Studies. He has published widely on energy issues, including *Producer–Consumer Dialogue: What Can Energy Ministers*

Say to One Another? (Chatham House, 2005); *The New Economy of Oil* (with Koji Morita, Norman Selley and Jonathan Stern, RIIA/Earthscan, 2001); *Companies in a World of Conflict* (editor, RIIA, 1998). John Mitchell retired in 1993 from British Petroleum, where his posts included Special Adviser to the Managing Directors, Regional Coordinator for BP's subsidiaries in the Western Hemisphere, non-executive director of various BP subsidiaries, and head of BP's Policy Review Unit. In 1976 he was an academic visitor in the Faculty of Economics at the University of Cambridge.

Index